earth resources

NOTRE DAME COLLEGE OF EDUCATION
MOUNT PLEASANT
LIVERPOOL L3 5SP

ENVIRONMENTAL STUDIES

Sig. Head of Dept...............

9. 76

THE PRENTICE-HALL FOUNDATIONS OF EARTH SCIENCE SERIES

A. Lee McAlester, Editor

STRUCTURE OF THE EARTH

S. P. Clark, Jr.

EARTH MATERIALS

W. G. Ernst

THE SURFACE OF THE EARTH

A. L. Bloom

EARTH RESOURCES

B. J. Skinner

GEOLOGIC TIME

D. L. Eicher

ANCIENT ENVIRONMENTS

L. F. Laporte

THE HISTORY OF THE EARTH'S CRUST*

A. L. McAlester and D. L. Eicher

THE HISTORY OF LIFE

A. L. McAlester

OCEANS

K. K. Turekian

MAN AND THE OCEAN

B. J. Skinner and K. K. Turekian

ATMOSPHERES

R. M. Goody and J. C. G. Walker

WEATHER

L. J. Battan

THE SOLAR SYSTEM*

J. A. Wood

*In preparation

earth
resources

second edition

BRIAN J. SKINNER
Yale University

PRENTICE-HALL, INC., *Englewood Cliffs, New Jersey*

Library of Congress Cataloging in Publication Data

SKINNER, BRIAN J. (date)
  Earth resources.

  (Foundations of earth science series)
  Bibliography: p. 143
  Includes index.
  1. Mines and mineral resources. 2. Natural
resources—United States. I. Title.
TN146.S54 1976   553   75-14156
ISBN 0-13-223016-X
ISBN 0-13-223008-9 pbk.

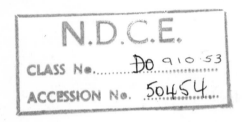
10 9 8 7 6 5 4 3 2 1

Printed in the United States of America

PRENTICE-HALL INTERNATIONAL, INC., *London*
PRENTICE-HALL OF AUSTRALIA, PTY. LTD., *Sydney*
PRENTICE-HALL OF CANADA, LTD., *Toronto*
PRENTICE-HALL OF INDIA PRIVATE LIMITED, *New Delhi*
PRENTICE-HALL OF JAPAN, INC., *Tokyo*
PRENTICE-HALL OF SOUTHEAST ASIA (PTE.) LTD., *Singapore*

contents

eight

nine

introduction

Our entire society rests upon—and is dependent upon—our water, our land, our forests, and our minerals. How we use these resources influences our health, security, economy, and well-being. (John F. Kennedy, Message on National Resources; Congress, February 23, 1961.)

Resource is a word with many shades of meaning. Dictionary definitions range from "something in reserve" to "additional stores, ready if needed." But the definitions do not specify the "somethings" and the "stores." They could be resources of courage to face a personal crisis, of wood to fuel a stove through a winter, or of finances to meet a medical expense. The resources discussed in this book are all linked by a common factor. They are all *natural resources,* which means that they are supplies we draw from a bountiful Earth, such as food, building and clothing materials, minerals, water, and energy.

All life needs and uses natural resources. Man alone has changed natural distributions and productivities, and by doing so systematically, has shaped the form of controlled living we call civilization. He has managed to combat climatic extremes and to increase vastly the Earth's yield of palatable foods above Nature's random growth. As a consequence, he has expanded his occupation of the globe to its farthest reaches and proliferated *Homo sapiens* far beyond the numbers that were once in stable balance with an unmanipulated Nature. Maintenance of the Earth's huge population is now totally dependent on continuing supplies of natural resources: fertilizers to increase crop yields, water to drink and to irrigate crops, metals to build machines, fuels to energize them, and a myriad other materials. Without continuing supplies civilized society must collapse and the population wither.

Natural resources fall into two distinct categories. Resources derived from living matter, such as food, clothing, and wood, are *renewable resources* because

1

they are replenished each growing season. Even if one season's crop is consumed, the next season brings a renewed larder. But *mineral resources* such as coal, oil, atomic energy, copper, iron, and fertilizers are not renewed each season. They are nonrenewable, one crop resources and the Earth's supplies are fixed. The kinds of mineral resources, their distribution, their quantities, the amounts we use, and our ever growing dependence on them are topics covered in this book. There are, of course, other questions involved in the use of mineral resources. As technologically advanced countries depend increasingly on supplies from less developed countries, there are questions concerning politics. Within producing countries there are questions concerning the use and misuse of land that is laid open during mining operations. Questions of the excessive dispersal of waste products—pollution—also seem inevitably to follow man's intensive use of mineral resources. Fascinating and important as these economic, social, and political questions are, they can only be mentioned and cannot, in a short volume such as this, be accorded the treatment they deserve.

Mineral resources have become essential ingredients for life—building blocks of society. But are they sufficient for a healthy future, and are they sufficiently accessible to allow easy exploitation? Empires have flourished repeatedly through history because of their control over rich and easily exploited mineral resources, but they withered with the same frequency as the riches expired. Are the world's remaining resources so distributed that the historical pattern of power dependence on resource availability is now a thing of the past, or is it still the key to the future? Questions such as these are very controversial and cannot yet be answered unambiguously. But the questions are posed and they must be answered by the generation now rising, the generation that will read this book. We all know that mineral resources are not uniformly distributed. It also seems apparent that some materials, such as oil, are so limited they cannot long be consumed at present rates. Inevitably, adjustments and changes will occur and they will influence all of us. Changes are already happening. But there are choices and each of us, directly and indirectly, will have our say. We should face these future choices with as much understanding as possible. This book does not provide all the answers, but it can help each of us in reaching our future decisions.

Study of the abundance and distribution of the Earth's resources is, in its more basic aspects, a branch of geology, and this book very properly belongs in an Earth Science Series. But study of resources is not just a branch of geology. It is a topic that involves each and every one of us. It is a starting point for study of the founding and history of civilization and of man's occupancy of the Earth. Hopefully, therefore, humanists will feel as comfortable as scientists with the pages of this book.

one

resources:
what and why

But the needed materials which can be recovered by known methods at reasonable cost from the earth's crust are limited, whereas their rates of exploitation and use obviously are not. (Walter R. Hibbard, Jr., in "Mineral Resources: Challenge or Threat?" Science, v. 160, p. 143, 1968.)

A healthy, hard-working person can produce just enough energy to keep a 100-watt light bulb burning. This may seem unimportant, but it is a humbling reminder that muscle power is really very puny. To our ancestors the limitations of the human muscle were all too evident, and they found it necessary to develop supplementary sources of energy. First they domesticated animals, but increasingly they learned to use more sophisticated means, such as sails, water wheels, wind mills, steam and internal combustion engines, and eventually electric motors.

Supplementary energy now exceeds muscle energy in every part of our lives from food production to recreation. It is like a gang of silent slaves who labor continually and uncomplainingly to feed, clothe, and maintain us. The energy comes, of course, from mineral resources such as coal, oil, and uranium, not from real slaves, but everyone on the Earth now has "energy slaves" working for him. In India the total supplementary energy produced is equivalent to the work of 15 slaves, each working an eight-hour day, for every man, woman, and child. In South America everyone has approximately 30 "energy slaves," in Japan 75, Russia 120, Europe 150, and in the United States and Canada a huge 300. The concept of "energy slaves" demonstrates how utterly dependent the world has become on mineral resources. If the "slaves" were to strike (which means if the supplies ran out), the world's peoples could not keep themselves alive and healthy. Reverting to muscle power alone would bring starvation, famine, and pestilence. Nature would quickly reduce the population.

It would be misleading to dwell on energy alone. The use of all natural resources is intertwined. Oil is of little use unless we have engines built of iron, copper, lead, zinc, and other metals. Farm lands will only yield maximum crops if they are tilled by tractors and plows and fertilized with compounds of phosphorus, nitrogen, and potassium. Figure 1-1 demonstrates dramatically the

FIG. 1-1 Increased food demands bring greater crop yields which in turn require increased inputs of mineral resources. In North America the average yield of corn per thousand square meters more than doubled between 1950 and 1970. Increased amounts of machinery, fertilizers, and energy were needed. The energy figures take into account such factors as the energy used in making tractors and in preparing fertilizers, insecticides, and herbicides. (After Pimental et al., 1973, *Science*, v. 182, p. 443.)

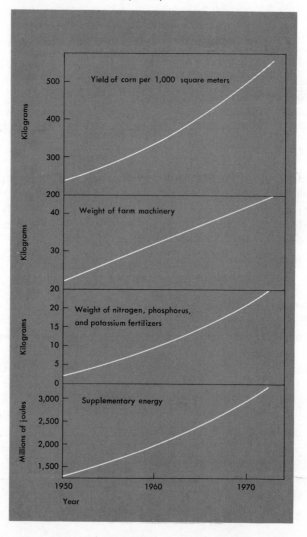

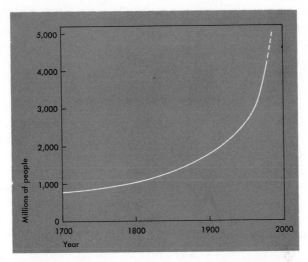

FIG. 1-2 Growth rate of the Earth's human population.

expenditure of mineral resources that underlies the huge yields of food needed to feed the world's massive population.

How far can food production be pushed? The question is debatable but nevertheless vital. Food is a renewable resource and if we were to total each crop, year after year, the sum would be endless and reach astronomic numbers. But the total amount of food that the Earth might eventually produce is only an academic nicety. What matters is how much food can be produced in any growing season. We cannot consume renewable resources faster than we produce them, and there is clearly a limit to the *rate* of food production. The limit is imposed by the size of the Earth, the amount of sunlight falling on it, and how efficiently plants can use the sunlight.

The population of the Earth is increasing dramatically (Fig. 1-2). Despite the so-called green revolution and great advances in farming methods, the population is increasing faster than the food production. Even with enormous efforts to expand production from all sources, including the oceans, it is inevitable that the population must eventually cease growing. A recent critical study of resources sponsored by the U.S. National Academy of Sciences estimated the Earth's theoretical carrying capacity at approximately 33 billion people— 10 times the world's present population. A measure of the severity of the resource limitation is that even if all possible means to increase food productivity were used, and strict regimentation of diet were practised, the 33 billion people could only be fed at a level of chronic starvation. Furthermore, the production rate would be a theoretical maximum, attainable for a short spurt, and, because inexpensive mineral resources would be needed to attain the goal, not possible on a continued basis. If we maintain the present birthrate, this doleful day will be reached about 100 years from now. Clearly, we should be foolish to reach such a disaster point, and a more realistically sized population, one which can live in equilibrium with a stable rate of food production, must be reached. The definition of this realistic size and the means of holding it in stable balance are the two greatest problems facing mankind today.

Balance and control of population will demand societal responsibility beyond any that mankind has ever attained. But they will not be faced until the controls that mineral resources exert on population size and on living standards are appreciated. Nor can the definition of a stable population proceed unless the Earth's supply limitations are accurately assessed. As the reader of this book will discover, assessment is a very complex process and remains to be completed. But relative abundances are already known and trends can be discerned. On the basis of such trends some experts suggest that the stable population size has already been exceeded. If they are correct, we should now be reducing our population size. Other experts suggest that the limit still lies ahead, but that numbers must level off before the present population size doubles.

There is a strong element of personal choice in all decisions regarding populations. The more people there are, the less there is to go around and the lower the living standard will be. Choice of life styles differ. Individual countries or regions will have to decide what their living standards and their stable populations will be, and in this way each of us will be involved in the most momentous decision his country has ever made. To commence our discussion of mineral resources, let us first consider what is meant when we say consumption rates are growing.

WHAT IS THE GROWTH RATE?

The consumption rate of all mineral resources is growing. We may tend to think of growth rates as linear processes, meaning that they grow by a constant amount each year. But the population growth curve in Fig. 1-2 is certainly not a straight line and is obviously not linear. Instead, it is growing by an increasing amount, or constant percentage of the whole, each year. The mathematical way to describe a curve such as Fig. 1-2 involves a symbol called an *exponent*. For example, in the expression X^3 the exponent 3 indicates that X is to be multiplied by itself twice. Thus, $10^3 = 10 \times 10 \times 10 = 1,000$. We therefore refer to curves such as Fig. 1-2 as *exponential curves*, and we refer to the growth they describe as *exponential growth*. Compound interest on a savings bank account is a familiar form of exponential growth. As any account holder knows well, the amount paid in interest is a little larger each year, and, after a certain number of years, the size of the account will double. The *doubling time* depends, of course, on the interest rate.

The population curve in Fig. 1-2 is growing at a rate of 2.1 percent a year and the doubling time is 33 years. If the rate remains constant, the world's population, which was about 3.5 billion in 1970, will rise to 7 billion by the year 2003, to 14 billion by the year 2036, and to 28 billion by 2069.

Consumption of mineral resources displays exponential growth and is increasing at a faster rate than population growth. Use of gold, for example, increases at about 4 percent a year, corresponding to a doubling period of 18 years, while consumption of fertilizers grows at nearly 7 percent for a doubling period of 10 years. When we examine consumption figures more closely, we find two components to the growth rate. The first is clearly the increasing population—more people need products to support them. The second component is an index of a rising standard of living around the world and the increasingly complex technological underpinning of our society. If we divide the annual gross consumption of a mineral product by a number equal to the population, we get the *per capita consumption*. For all mineral products the per capita consumption curves are also increasing exponentially. Consumption curves, therefore, are the sum of the two separate exponential curves, and the population problem is a double-barreled one. Not only must the size eventually become stabilized, so too must living standards. When that time is reached, one unfortunate consequence of exponential consumption curves can be eased. The unfortunate consequence is that recycling can never satisfy the demands of exponential growth. No matter how efficient the recycling may be, new material must be continually added from the Earth's dwindling stock.

To be sure that we are seeing an exponential growth rate, we should conduct observations over a number of years because short-term fluctuations occur, such as those caused by wars and depressions. Long-term observations over time spans of 50 years or more all point to the story illustrated in Fig. 1-3: Despite fluctuations, consumption of mineral resources continues relentlessly upward, so that by 1974 the per capita consumptions in the United States were sand and gravel, 4.1 metric tons; crushed stone, 3.9 metric tons; iron, 0.6 metric tons; petroleum, 3.8 metric tons; coal, 2.4 metric tons. These figures highlight the most direful consequence of exponential growth. When supplies are fixed in size, demand must ultimately exceed the supply, no matter how large the supply may be. Consider an absurd extreme. Throughout the present century, the world's annual production of mineral resources has doubled every 10 years. In 1973, the world production of all kinds of new mineral resources was an estimated 15 billion metric tons. If exponential growth were maintained to the year 2213—that is, for 240 years—the people then living would have to produce 250,000,000 billion metric tons a year. This is absurd indeed; that mass is equal to all of the land standing above sea level.

Exponential growth cannot continue forever, and we must ask ourselves what future consumption curves might look like. Some experts believe they can detect a flattening of consumption curves in countries with high living standards and from this interpretation they predict an eventual leveling off to a constant consumption rate. While possibly correct locally, world consumption of mineral resources is still growing exponentially, and because industrialization

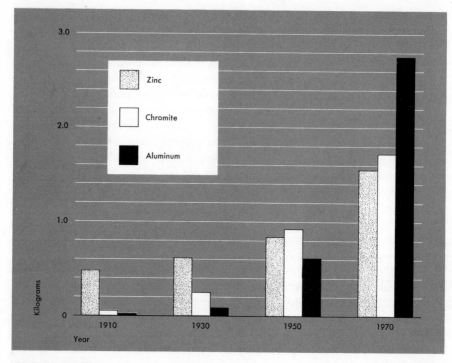

FIG. 1-3 Increasing per capita consumption, on a worldwide basis, of some representative mineral products. Zinc metal is used in alloys and as a corrosion-resistant coating. Chromite is a source of metallic chromium, a vital alloying metal in steels, and is also used for high temperature furnace linings and in preparing chemicals. Aluminum is used principally in the building, transportation, packaging, and electrical industries. (After U.S. Bureau of Mines.)

FIG. 1-4 Per capita consumption of aluminum on different continents, 1970.

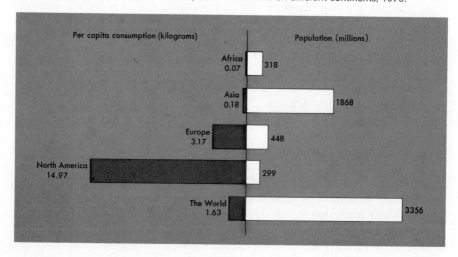

and living standards are very unevenly developed around the world, the geographic consumption of mineral resources is not uniform. As an example, consider the very uneven per capita consumption of aluminum in the year 1970 (Fig. 1-4), and note the low consumption in Asia where half the world's population resides. Simply raising the world's living standards and the per capita consumption of all mineral resources to the present rate enjoyed by the inhabitants of North America and Europe would severely tax the known reserves of many metals.

The geographic distribution of mineral resources on a world-wide basis is very uneven. Countries may be rich in one commodity but poor in another: No technically advanced country is currently self-sufficient and it seems unlikely that any country ever can be self-sufficient. The United States, for example, is self-sufficient in very few commodities and for some, such as platinum and chromium, it must import all that it consumes (Table 1-1). Production and

Table 1-1   Selected U.S. Mineral Commodities Consumed During 1972 That Had To Be Imported

Mineral Commodity	Percentages Imported	Main Supply Countries
PLATINUM	100	South Africa, U.S.S.R., Canada, Japan, Norway
CHROMIUM	100	U.S.S.R., South Africa, Turkey
TANTALUM	97	Nigeria, Canada, Zaire
ALUMINUM ORE	96	Jamaica, Surinam, Canada, Australia
MANGANESE	95	Brazil, Gabon, South Africa, Zaire
ASBESTOS	85	Canada, South Africa
TIN	77	Malaysia, Thailand, Bolivia
NICKEL	74	Canada, Norway
GOLD	61	Canada, South Africa, U.S.S.R.
SILVER	44	Canada, Peru, Mexico, Honduras, Australia
PETROLEUM	29	Central America, Canada, Middle Eastern countries
COPPER	18	Canada, Peru, Chile

(From "Minerals and Mining Policy, 1973," a report by the Secretary of the Interior to the U.S. Congress.)

consumption of mineral resources, therefore, raise complex questions in international trade and politics. If we are confidently to face both aspirations for high standards of living and maintenance of a large population, it is imperative that we appreciate and understand the nature and distribution of resources we must use.

KINDS OF MINERAL RESOURCES

Throughout this book we shall be discussing mineral resources in their most general sense, and the term will be taken to include all nonliving, naturally occurring substances that are useful to man whether they are inorganic or organic. Thus all natural solids, fossil fuels such as petroleum and natural gas, as well as the waters of the Earth and gases of the atmosphere, fall under this definition of mineral resources. We shall classify the different types of mineral resources on the basis of use, as demonstrated in Fig. 1-5.

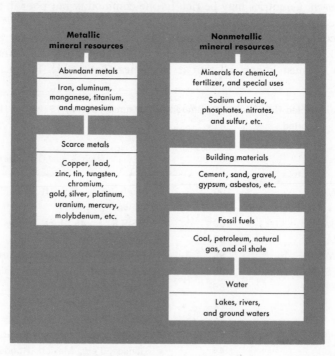

FIG. 1-5 Kinds of mineral resources.

It is instructive to compare the relative values we place on different kinds of mineral resources. The value of mineral resources used in the United States, divided into fuels, metals, and nonmetals (excluding water), is plotted in Fig. 1-6. Fuels, accounting for 65 percent of the total value, are the clear leaders. Similar relative values of minerals consumed pertain in all industrial countries. Figure 1-6 demonstrates two important relations discussed earlier. First, the overall value is increasing exponentially, reflecting exponential consumption rates. Although Fig. 1-6 is not corrected for a declining value of the dollar, exponential growth in consumption is observed no matter what correction factor is used. Second, increases in metals and nonmetals are parallel to the increase in fuels, reflecting the interdependence of all mineral resources.

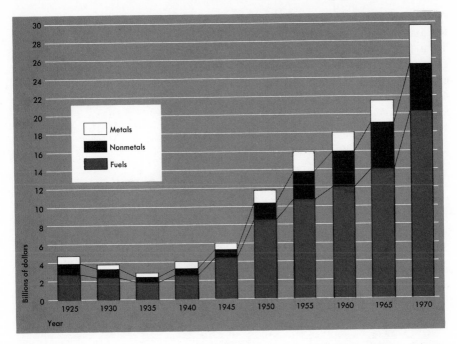

FIG. 1-6 Total value of mineral resources produced annually in the United States. Note the decline that occurred during the depression years of the 1930's. This illustrates why it is necessary to consider long periods of time when determining trends in mineral resource consumption. (After U.S. Bureau of Mines.)

A WORD ABOUT MEASUREMENT UNITS

We are all familiar with different ways of measuring volumes. For example, some things are commonly measured by the cubic foot, others by the cubic centimeter; many liquid volumes are measured by the gallon or liter, wood is measured by the cord and the board foot, fruit and vegetables by the bushel, and oil and cement by the strangest unit of all, the barrel. A barrel of oil is equal to 42 U.S. gallons, while a barrel of cement is equal to 170.5 kilograms. The same kind of confusing range occurs among units of weight, length, and energy, and in no field is the confusion more widespread than in mineral resources. Fortunately, units are slowly being changed and standardized, but most mineral resources are still measured in archaic units that are retained by custom rather than by reason.

To avoid the inconvenience of continually converting from one unit to another, the units used throughout this volume will be metric units. The units of length are the centimeter, meter, and kilometer; of volume, the cubic centimeter, cubic kilometer, and liter; of mass, the gram, kilogram, and metric ton; and of energy, the joule. Conversion factors between these basic units and those commonly used in commerce and industry are given in Table A-1 of the

Appendix. Perhaps the least familiar units for most readers are the metric ton and the joule. A metric ton is 1,000 kilograms and is equal to 2,204 pounds. To a first approximation, therefore, a metric ton can be equated with the common long ton of 2,240 pounds. A joule is a unit of energy used mainly in electricity, and it is equal to 0.239 calorie. To a first approximation, therefore, 4 joules can be considered equivalent to a calorie.

two

where are resources found?

... *minerals are not renewable. They are formed in the earth's crust by infinitesimally slow natural geologic processes acting for thousands or millions of years. (D. A. Brobst and W. P. Pratt, U.S. Geological Survey, Professional Paper 820, 1973.)*

RESOURCES VERSUS RESERVES

When a new copper mine is opened or a new oil field discovered, one of the first questions asked is always, how much does it contain? Similarly, when the future of a mine is discussed, we ask, how much is left? But how much of what? Do we mean material that can be profitably recovered today, or do we mean material that can only be recovered some time in the future when prices are higher, or when a new recovery method has been invented? And how closely must the answer be estimated? To the nearest kilogram, to within 5 percent of the total, perhaps even to within 50 percent of the amount present? These are essential questions for all mineral resources. Before facing them, a few clarifying points of terminology will help.

A mineral resource is any presently or potentially extractable concentration of naturally occurring solid, liquid, or gaseous material. Extraction must of course be profitable, so note the distinction drawn between presently and potentially extractable materials. The term mineral resource has two components. The first, or presently extractable component, is an *identified resource*, or, as it is more commonly called, a *reserve*. We know how to exploit reserves, where to find them, and the cost of recovery. We can rely on them and they can be profitably recovered today. The second, or potentially extractable component, is a *potential resource* that still has many ifs, ands, or buts attached to it. Perhaps new mining technology must be developed or new processing methods found, or perhaps government price supports are needed. Whatever the reasons, unanswered questions surround potential resources. Future developments are needed before they can become reserves. But potential resources are materials that deserve special attention and which for reasons of size, richness, or location lead us to postulate that they might one day be used and become reserves. The certainty with which we can guess the likelihood of potential resources ever becoming reserves decreases the further we move from common experience, the more unconventional the materials we include in the estimate, and the further we project our guesses of economic and technological changes into the future. One point that seems clear, however, is that man will always mine selectively, seeking out the richest concentrations first. The suggestion occasionally made that we will soon turn to common rocks, essentially ignoring the local concentrations we commonly call ore deposits, seems highly improbable.

THE EARTH

Resources are everywhere—in the atmosphere, the oceans, high in the mountains, in rocks beneath the sea floor, and indeed in every accessible region of the Earth—but they are not uniformly distributed. To the contrary, they are very unevenly distributed and they differ in kind and amount and from place to place. In order to put the distribution of resources in perspective, then, we should briefly examine some of the essential features of the planet on which we live. Readers wishing to obtain a more detailed picture should consult other volumes in this series, and in particular, *Structure of the Earth* by S. P. Clark, Jr., *Oceans* by K. K. Turekian, *Man and the Ocean* by B. J. Skinner and K. K. Turekian, and *Atmospheres* by R. M. Goody and J. C. G. Walker.

Earth has a mass of 6.5×10^{21} metric tons and is comprised of 88 different chemical elements; the total tonnage of any element is truly enormous, even for those present at extremely low concentration levels. Because most of the Earth is inaccessible, the tonnage actually within reach of the Earth's surface is millions of times less than the Earth's total mass. At the Earth's center is a metallic *core* consisting predominantly of iron and nickel, surrounded by a *mantle* of dense rock rich in iron and magnesium; the core and mantle together account for more than 99.6 percent of the total mass of the Earth. Above the mantle is the Earth's *crust*, which is the only portion of the solid Earth we actually observe, and which accounts for 0.375 percent of the Earth's mass. The crust is of two parts, one that projects above the oceans and one that lies below; the portion above, in addition to a narrow sea-covered fringe around each continent, is called the *continental crust*; the portion below the oceans is the *oceanic crust*.

At the Earth's surface are the oceans, lakes, and rivers that, together with the water trapped in holes and fractures in soil and near-surface rocks, are called the *hydrosphere*, and account for 0.025 percent of the Earth's mass.

Enclosing everything is the gaseous envelope of the *atmosphere*, which accounts for only 0.0001 percent of the mass. It is from the three outermost, and smallest, zones—the crust, hydrosphere, and atmosphere—that we draw our present resources and to which we must look for those of the future. The mantle and the core are so inaccessible that they cannot ever be seriously considered as potential suppliers of resources.

The Crust

Most mineral resources are derived from the crust; we will therefore consider it first. The crust has obvious and important differences from the hydrosphere and atmosphere. First, it is predominantly composed of *minerals* that are crystalline solids with specific and rather simple compositions. Second, as any walk through a rocky terrain will reveal, minerals are not randomly distributed, but are relatively concentrated into specific and distinctive groupings called *rocks*. Limestone, for example, is a rock consisting mostly of the mineral calcite, $CaCO_3$; quartzite is mostly quartz, SiO_2; and coal is a rock composed of solid organic matter. The chemical elements are therefore not evenly distributed through the crust but are distinctly segregated. In this fashion, even elements that have a low *average* concentration in the crust are sometimes found in exceedingly high *local* concentrations. The richest local concentrations are ore deposits.

Since it is difficult to sample the floor of the ocean, the composition of the oceanic crust is not known with complete certainty. There are indications, however, that it is relatively constant in composition, consisting largely of minerals rich in calcium, magnesium, iron, and silicon, and that it formed by the cooling of lavas extruded on the sea floor to form a type of rock called *basalt*. The oceanic crust seems to be truly a submarine phenomenon. Rarely, and only in special circumstances, is it elevated above sea level and subjected to the same forces of erosion that wear down the continents.

The continental crust, which is somewhat more than half the mass of the whole crust, or approximately 0.29 percent of the Earth's mass, contains less iron and magnesium than the oceanic crust, but relatively more silicon, aluminum, sodium, and potassium. The entire crust, oceanic and continental, plus a relatively cool upper portion of the mantle underlying both segments seems to float or be buoyed up on the rest of the mantle. Indeed, the outer 100 kilometers of the entire Earth is apparently floating and slowly moving, like a series of great rafts or plates, over the mantle. The floating property helps explain why ocean basins are low and continents are elevated. Rocks of the oceanic crust are more dense than those of the continental crust. The continents float high, like blocks of light wood floating on water; oceanic crust sits lower, like blocks of heavy wood. Because water seeks the lowest level, it is in the low-lying basins created by the heavy oceanic crust that water of the hydrosphere has concentrated. There is, however, a little more water in the hydrosphere than the basins can contain. The oceans therefore spill over onto the continental margins, creating a submerged continental shelf and continental slope (Fig. 2-1). As we shall see, these submerged margins of continental crust have a potentially important role in the future of mineral resources.

A systematic examination of all known rock types shows that two principal kinds predominate. The first are *igneous rocks*, formed by the cooling and crystallization of liquids from deep in the crust or upper part of the mantle, called *magmas*. The second are *sedimentary rocks*, formed by compaction and cementation of sediment derived from the continuous erosion of the continents by water, atmosphere, ice, and wind. Most of the sediments are deposited in the sea along the margins of continents. As the marginal piles of sediment grow larger and are buried deeper, increasing pressure and rising temperature produce physical and chemical changes in them. The resulting *metamorphic rocks*, however, generally show whether they were originally sedimentary or igneous rocks. When a sedimentary pile becomes thick enough, material near the bottom may melt to form magma. The newly formed magma, being less dense than the rocks from which it was derived, will tend to rise up, intruding its parents and as it cools and crystallizes it will form a new igneous rock. Thus, there seems to be a sequential process in rocks, whereby igneous rocks become sediments, sediments become sedimentary rocks, then metamorphic

FIG. 2-1 Diagram of the edge of the South American continent off Argentina. The edge of the continental crust is marked by the steep continental slope. Sea water overlaps part of the continent to form the continental shelf. (After B. C. Heezen and M. Tharp, for the Geological Society of America.)

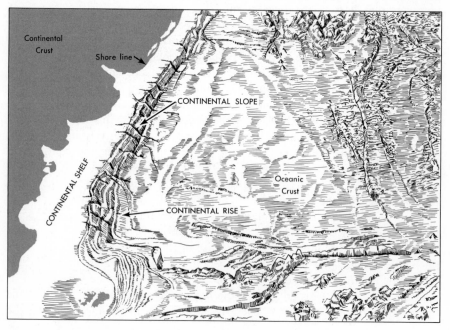

rocks, and eventually igneous rocks again. The sequence is slow—hundreds of millions of years are necessary—and the details are complicated. As weathering and erosion occur, some substances are dissolved and removed in solution while others are transported as suspended particles. Extreme chemical separation can therefore occur and many of our mineral resources form as a result. It is a fortunate consequence of planetary evolution that the atmosphere and hydrosphere, so necessary for life on the Earth, have also been the essential media which produced the necessary chemical separation to form most of the rich mineral resources.

Continental crust contains an extremely varied array of rock types. It also contains a great deal of evidence to suggest that the rock forming processes we can observe today have been active for at least 3,500 million years. Apparently, however, the atmosphere and waters of the hydrosphere have not always had the same composition because the frequency of certain types of rocks and mineral resources has changed through the ages.

The oceanic crust, by contrast with the continental crust, shows little variation in composition. This leads to a suspicion that the rocks of the sea floor might not contain as many valuable mineral resources as do the rocks of the continental crust. Resolution of this suspicion will be one of the prime goals of oceanographic research in the decades ahead.

Scientists have long sought a single explanation for such diverse features as the compositional differences between oceanic and continental crust, the near parallelism of coast lines on either side of the Atlantic Ocean, and the topography of mountain ranges, both on the ocean floor and on the land. An end to the search came during the 1960's when it was discovered that the outer 100 kilometers of the solid Earth slides slowly around like a loose skin. The sliding skin carries the continents with it and rather than being a single, coherent sheet, it is broken into six huge fragments, or plates, plus a number of smaller ones. New igneous rock that forms continually along giant fracture zones rending the ocean bottoms tends to push the plates sideways. The moving plates eventually bend and plunge down into the mantle again so that each plate acts somewhat like a conveyor belt. Although details are poorly understood, the distribution of many mineral deposits seems to be somehow related to the present and past fractures that bound the moving plates, and many scientists believe that a vastly improved understanding of how and why mineral deposits form will be forthcoming during the remainder of the twentieth century.

The average composition of the continental crust (Fig. 2-2) reveals that only 9 elements account for 99 percent of the mass. There are still 79 elements to be accounted for, and as they total 1 percent of the crust, they can only be present in trace amounts. Many of the elements that are vital resources fall in this category. Fortunately, there are a number of special ways in which local concentrations, or ores, of the scarcer elements have occurred within the crust. Before discussing how these concentrations formed, and how large they are, let us look briefly at the two outer zones of the Earth.

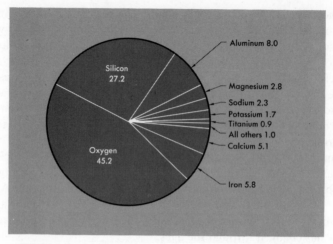

FIG. 2-2 Major elements in the continental crust, expressed as percentages. (After K. K. Turekian, 1969)

The Hydrosphere

The hydrosphere is the entire body of water on and near the surface of the Earth. It is, to some extent, intermingled with the crust and to an important degree, the crust and hydrosphere react with each other to cause weathering, the saltiness of the sea, and many other common phenomena. Water itself is the most valuable resource of the hydrosphere and will be discussed separately in Chapter 9.

The hydrosphere is the site of other important resources besides water.* The oceans, which cover 70.8 percent of the Earth's surface to an average depth of 3.96 kilometers, act as a collection reservoir for many of the soluble materials formed on the Earth, such as those released from rocks and soils by weathering and those given off in volcanic gases. Through the ages they have reached their present composition level of 3.5 percent dissolved solids by weight in solution. A dissolved solid content of 3.5 percent amounts to approximately 38×10^6 metric tons of dissolved matter in every cubic kilometer of sea water, and it has been shown that the relative proportions of the major elements in solution are essentially constant throughout the oceans of the world. Sodium and chlorine, the two elements present in common salt, are by far the most abundant dissolved elements, and these, together with magnesium, sulfur, calcium, and potassium, account for 99.5 percent of all dissolved solids in the sea (Fig. 2-3). Each cubic kilometer of sea water contains significant amounts of 64 other elements such

---

* For a more complete discussion of resources of the sea, readers are referred to a companion volume, *Man and the Ocean* by B. J. Skinner and K. K. Turekian.

as zinc and copper, each 2,000 kilograms; tin, 800 kilograms; silver, 280 kilograms; and gold, 11 kilograms.

Despite the wide range of elements present in sea water, only four are being commercially recovered in significant quantities at present: sodium and chlorine (recovered jointly in the form of common salt), magnesium, and bromine. Sodium chloride (see Chapter 7) is locally produced by solar evaporation of sea water in shallow ponds. Bromine is recovered either by the addition of the complex organic compound aniline, which causes the precipitation of insoluble tribromoaniline, or by the addition of chlorine gas, which causes bromine gas to be released. Magnesium (see Chapter 5) is recovered by the addition of dissolved calcium hydroxide causing the precipitation of magnesium hydroxide.

The composition of sea water with respect to its abundant elements is well established, and the volume of the ocean basins measured accurately. The sea's resources of the elements we now recover must therefore be classified as reserves. The reserves are so enormous that they will fulfill every conceivable demand. The potential resources of the other elements, including those present in trace amounts, but which are not presently recovered, are also known with a high degree of certainty. Because the volume of the seas is $1,375 \times 10^6$ cubic kilometers, the potential resources are truly enormous, and one must question why man has not exploited them more extensively. Two difficulties are obvious. Unless specific reactions can be found to remove only the element or elements of interest, as in the removal of bromine by precipitation of tribromoaniline, all the other dissolved compounds must also be removed—a process that consumes a vast amount of energy and is wasteful unless uses can be found for the

FIG. 2-3 Average composition of dissolved salts in sea water on a weight percentage basis.

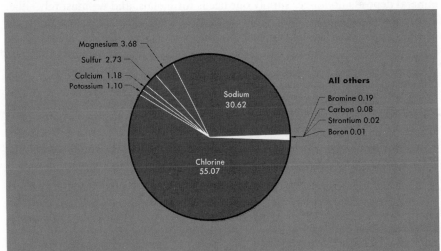

large amounts of materials so produced. Furthermore, although the total amounts of minor elements present in the sea are large, the solution is still a very dilute one, and exceedingly large amounts of water must be processed in order to recover the small quantities dissolved.

It is interesting to compare the average composition of the continental crust and of sea water. The valuable elements, such as gold and copper, are much more abundant in the crust (Table 2-1). If we ever see the day when identifiable

Table 2-1 Valuable Metals in a Cubic Kilometer of Average Continental Crust and Sea Water

Metal	Amount in Continental Crust (metric tons)	Amount in Sea Water (metric tons)
MANGANESE	1,809,000	1.9
ZINC	170,000	2.0
CHROMIUM	130,000	0.2
NICKEL	100,000	2.0
COPPER	86,000	2.0
COBALT	32,000	0.05
URANIUM	7,800	3.3
TIN	5,700	0.8
SILVER	160	0.3
GOLD	5	0.01

mineral deposits no longer supply our needs for metals, and we must rely either on average rocks or sea water, Table 2-1 suggests the odds are strongly in favor of common rocks. The odds could, of course, be tipped in favor of sea water if an extraction process were at least a thousand or more times more efficient than a comparable extraction from rocks. Unfortunately, the extreme dilution of many elements in sea water makes this difficult because of the sheer volume of water that must be handled. It has been recently estimated, for example, that if sea water must be pumped during processing, and if a plant capable of handling a volume as great as 4.5 million liters a minute works with a 100 percent extraction efficiency, the cost of pumping and processing would exceed the value of the extracted material for all elements less abundant than boron. If pumping could be avoided, costs would clearly be much lower. One possible pump-free extraction process has already been proposed. Sea water contains approximately 3.3 metric tons of uranium per cubic kilometer. This can be extracted when sea water is passed over titanium hydroxide coated on fibers of glass. An intriguing suggestion is that pumping could be avoided if bags of such fibers could be exposed to tidal currents, or used as linings on the bottoms of ships. The process has not yet been tried outside the laboratory, but it does hold some hope for the future. Similarly, innovative new chemical extraction methods might be found for other elements; the rewards, like the problems, would be enormous. Unfortunately, most authorities view the chances of success as extremely remote. We must conclude, therefore, that for sodium, potassium, magnesium, calcium, and strontium; for the halogen elements chlorine, bromine, iodine, and fluorine; and for sulfur, boron, and phosphorus, the sea contains vast potential resources that may some day be exploited. For a large number of other elements the sea contains vast amounts, but they are unlikely ever to be exploited because more favorable sources of the same elements can be found in rocks.

The limited possibilities for elements from sea water should not divert our attention from the possibility of mineral resources contained in rocks on the floor of the ocean, or from energy sources from sea water, and both possibilities are discussed later in the book.

The Atmosphere

The atmosphere is continually mixed and has an essentially uniform composition; it is also accessible and easy to sample. The atmospheric composition is relatively simple and is known with considerable accuracy, so our estimates of the abundances of the various atmospheric gases fall in the category of reserves. Three gases—nitrogen, oxygen, and argon—account for 99.9 percent of the atmospheric volume, with nitrogen, an essential plant fertilizer element, the most abundant constituent (see Chapter 7).

Oxygen and argon, together with the rarer gases, neon, xenon, and krypton, are also recovered from the atmosphere, but in relatively small amounts. The uses to which gases are put do not permanently remove them from the atmosphere; thus, we can classify them as renewable resources. Also, because of the small amounts used, their temporary removal has no observable effect on the atmospheric composition.

For the few recoverable elements concentrated in it, the atmosphere provides an essentially limitless source.

three

energy
from fossil fuels

When consideration is given to the factual data pertaining to both the world and the U.S. rates of production of coal and oil ... two results of outstanding significance become obvious. The first of these is the extreme brevity of the time during which most of these developments have occurred. For example, although coal has been mined for about 800 years, one-half of the coal produced during the period has been mined during the last 31 years. Half of the world's cumulative production of petroleum has occurred during the 12-year period since 1956.
The second obvious conclusion ... is that the steady rates of growth sustained during a period of several decades ... cannot be maintained for much longer periods of time. (M. King Hubbert, "Energy Resources," in Resources and Man, *edited by P. Cloud, 1969, U.S. National Academy of Sciences.)*

We saw in Chapter 1 that supplemental energy resources are our most valuable mineral commodities. They are, therefore, the first resources to demand our attention.

Units and Sources of Energy

Before proceeding let us clarify terms. Energy means the capacity to do work. Units of measurement reflect the kind of work being done. For example, the *calorie* is defined as the heat energy needed to raise the temperature of 1 gram of water by 1°C; the *joule* is the electrical energy needed to maintain a current of 1 ampere for 1 second at a potential of 1 volt. Energy can, of course, be converted from one form to another, so the units can be interchanged; 1 joule is equal to 0.239 calories. There are many energy units in common use, such as the British thermal unit (Btu) and the erg, but to avoid confusion we will use only the joule.

Although the total energy available is an important number, we must also be concerned with the rate at which energy is used, and this means that a

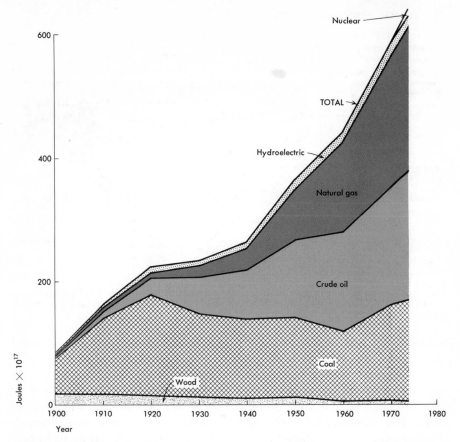

FIG. 3-1 Supplementary energy in the United States is produced from a number of different sources. Note that the total energy produced is currently increasing with a doubling period of about 22 years. (Data from U.S. Bureau of Mines.)

rate, or time-dependent property, must be introduced. This property is called *power* and is defined as energy used per unit time. Horsepower is one of the most familiar power units, but to maintain consistency, the power unit used in this book is the *watt*, which is defined as 1 joule of energy being consumed every second. The watt is an important unit for discussions of renewable energy resources, such as heat from the Sun. As with food and other renewable resources, the total amount of solar energy that has reached, and will reach, the Earth is not important. The important figure is the rate at which it reaches us.

Man's Use of Energy

The supplementary energy used by man, from all sources, is now about 3×10^{20} joules per year. Although this number is enormous, it is small by comparison with the solar energy received by the Earth each day: 1.5×10^{22} joules. A small amount of man's supplementary energy comes from renewable sources—hydroelectric schemes, wood burning, wind and water wheels—but by far the greatest amount comes from nonrenewable sources, and particularly the so-called fossil fuels, coal, oil, and natural gas (Fig. 3-1). It is interesting to

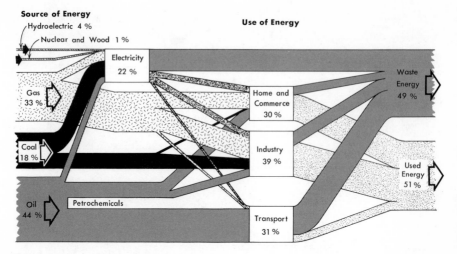

FIG. 3-2 Flow and use of supplementary energy in the United States during 1972, expressed as a percentage of the total energy use. Some oil and gas is used for nonenergy purposes, principally for production of petrochemicals.

observe from Fig. 3-1 that while coal and wood have not shown exponential growth rates in recent times, the overall use of energy clearly is growing exponentially. Figure 3-1 is drawn from data available in the United States, but similar kinds of growth patterns are observed in all countries.

An essential ingredient in discussions of natural resources is how we use them. This is particularly important where energy resources are concerned. The total energy use is important, but the source of energy has an important control on the kind of uses to which it is put. For example, expanding nuclear power stations will produce more electricity, but this will not help the truck driver who needs gasoline. The use of energy in an industrialized community such as the United States (Fig. 3-2) raises many interesting points. One is the relatively low efficiency of energy use. Poor insulation, inefficient engines, and similar wastes, however, are only part of the problem. There is an inherent loss whenever energy is stored, transported, or converted from one form to another. Despite this limitation by laws of nature, much of the lost energy does arise from our own inefficiency. The heavy dependence on fossil fuels as energy sources is evident from both Figs. 3-1 and 3-2, so they will be the topic with which we will commence our discussion of energy sources.

FOSSIL FUELS

The term fossil fuel refers to trapped remains of plants and animals in sedimentary rock. Although principally plant material, fossil fuels occur in many ways, depending on the kind of sediment, the nature of the original organic compounds, and the changes that have occurred through long geological ages.

Living plants trap energy from the Sun by the process of photosynthesis, and they store the energy in their chemical compounds. Most of the energy is released again when the plants die and decay. In many sediments, however, organic matter is trapped and buried before it decays completely. In this way some of the solar energy becomes trapped in rocks—hence the term fossil fuel. The rate of decay is very nearly equal to the rate of growth, so the fraction of organic matter trapped in any one growing season is tiny. But the accumulated remains over the 600 million years that life has been prolifically distributed across the Earth is now considerable. Because the accumulation rate is so slow—millions of times slower than we now dig up the organic matter and burn it—we must consider the fossil fuels as nonrenewable resources. The "big three" among the fossil fuels, coal, crude oil, and natural gas, now supply over 95 percent of the supplementary energy.

Coal

Coal forms from the remains of freshwater plants. Into densely vegetated, stagnant swamps fall dead limbs, trunks, leaves, and spores, soon to be water-logged and to sink. Once water-covered and protected from the atmosphere, bacterial digestion commences, turning the woody plant remains to a jelly-like mass of peat. But oxygen supplies in the water are quickly consumed, the bacteria die, and decay ceases. Thick accumulations can form only if the swamp basin slowly subsides during accumulation and rich deposits only occur when the inflow of mud and other debris is low. Coal accumulation sites are not widespread at present, but a typical one occurs in the Dismal Swamp of Virginia and North Carolina, where an average of 2 meters of modern peat covers a 5,700 square kilometer area.

The earliest fossils of land and freshwater plants are in rocks about 410 million years old. The oldest coal deposits occur in 370 million year old rocks and the most ancient coals of significant size, in the Canadian Arctic, are about 350 million years old. From 350 million to about 250 million years ago, the greatest coal forming period in the Earth's history occurred. Coal deposits were formed on all continents but by far the greatest deposits were then laid down in North America, Europe, and Asia. Subsequent periods of coal formation have occurred, but none has been so extensive nor so prolific as the great coal age that ended 250 million years ago.

Peat, the first stage in the formation of coal, is a low-rank material, which is to say that it has a relatively low carbon content and a low heat-producing or calorific value. Upon burial and compaction of peat, a series of reactions occurs, and much of the water, oxygen, nitrogen, and other plant elements originally present are expelled, leaving an increasingly dense and carbon-rich coal. The process of coalification proceeds with age, bringing an increase in rank (Fig. 3-3), so that older and more deeply buried coals generally have higher ranks than younger or shallower ones.

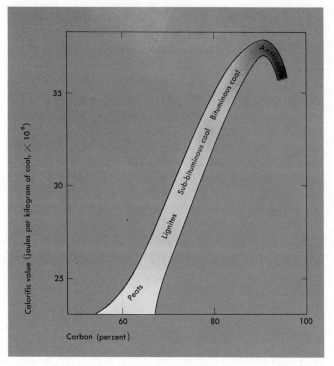

FIG. 3-3 Increasing calorific value of coal with increasing rank.

Coal basins seem to be remarkably stable and not subjected to meta-morphism and rapid erosion. Geologists believe that most of the coal ever formed is still on the Earth. They also believe that most of the major coal basins have already been discovered. On this basis, P. Averitt of the U.S. Geological Survey in 1969 estimated the world's reserves of all forms of coal to be 8,620 billion metric tons, and the potential resources to be 6,650 billion metric tons. Averitt considered recoverable such coal as occurred in seams greater than 30 centimeters thick, lying no deeper than 2,000 meters. In the case of coal it is reasonable to consider potential resources as being almost reserves. The term potential resources is used, for example, in the case of extension of known seams below present mining depths. There is, therefore, some justification for combining reserves and potential resources for a total of 15,270 billion metric tons, equivalent in heat energy to about 4.2×10^{23} joules. Unfortunately, it is impossible to mine all the coal from a seam, and in most cases of underground mining, a recovery of 50 percent is considered excellent. On this basis, the world's recoverable coal reserves amount to 7,135 billion metric tons, or about 2.1×10^{23} joules. As seen in Fig. 3-4, the geographic distribution of coal reserves is highly erratic.

Use of coal as a widespread fuel began in the twelfth century A.D., when inhabitants of the northeast coast of England found that inflammable black rocks weathering out of coastal cliffs were good substitutes for their rapidly disappearing forest woods. Known as "sea coles," the new fuel soon became widely used—to the extent that by 1273 outraged Londoners complained of repugnant odors and air pollution arising from coal-burning. This deterred no one, however, and the use of coal as a fuel spread rapidly. Although coal is no longer the predominant source of energy in the world, it remains predominant in some countries, and limitations on supplies of oil and gas suggest that use of coal might everywhere increase rapidly in the decades ahead. By the year 2000 coal may again be the predominant fossil fuel used in the world.

Desirable as coal may be for a fuel, many difficulties attend its use. All coal contains from 0.2 to about 7.0 percent sulfur, present as the iron sulfide mineral, pyrite, FeS_2, as ferrous sulfate, $FeSO_4 \cdot 7H_2O$, or gypsum, $CaSO_4 \cdot 2H_2O$. When coal is burned, the sulfur is released to the atmosphere causing environmental hazards. Low sulfur coal can be burned without serious consequences, but most coal will have to be processed prior to burning in order to reduce the level of sulfur. Treatment costs may be high and concerns have been expressed by many experts for reliability of extraction. The sulfur odors that were annoying Londoners in 1273 are proof that the problem is a long-standing one still waiting for technology to supply a solution. Another problem of vital concern is the mining of coal. Extraction by underground methods is unpleasant, inherently dangerous, and seems to carry the long-term threat of unavoidable health

FIG. 3-4 Geographic distribution of coal reserves. More than 97 percent of the recoverable coal occurs in North America, Europe, and Asia, leaving the continents of the Southern Hemisphere relatively coal-poor. (After P. Averitt, 1969.)

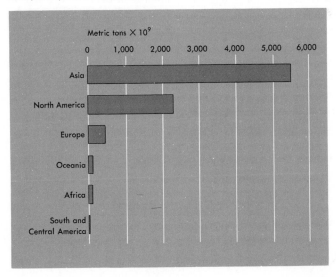

hazards. Surface mining, though more efficient and less dangerous than underground mining, causes serious disruptions of the land surface. Furthermore, only about 4 percent of all coal lies close enough to the surface or occurs in thick enough seams to be recovered by strip mining. For the United States this amounts to some 130 billion metric tons out of a total of 3,900 billion metric tons. In the long run, therefore, underground extraction of coal, despite its many difficulties, will presumably be the predominant recovery method.

Crude Oil and Natural Gas

Crude oil and natural gas, the liquid and gaseous components of petroleum, occur together, contain many of the same compounds, and have common origins concerning which there is still some controversy. Petroleum is a derived material and has the characteristic of migrating from its site of formation; the precursors are therefore not always readily identified.

Crude oil and natural gas are composed chiefly of hydrocarbons and, like coal, are found in sedimentary rocks (Fig. 3-5). Although not confined to marine sediments, petroleum is more abundant there than in freshwater sediments, and it is conspicuously more abundant in sedimentary basins with a high percentage of organic-rich sediments. Almost all sediments contain some organic debris, however, and a wide variety of petroleum hydrocarbons and even droplets of crude oil have been found associated with little altered organic matter in many common rocks such as shales and limestones. There is little doubt, therefore, that widespread sedimentary organic matter, of microscopic plant and animal origin, is the source of petroleum; and further, it would seem that petroleum formation begins immediately after burial of the organic matter.

The earliest formed petroleum compounds tend to have high molecular weights, like those in the living cells from which they are derived, and produce very viscous oils. On burial, and as the temperature and pressure rise, the large molecules are broken or *cracked* into lighter and more mobile ones. The longer the process continues, the "lighter" the crude oil becomes. Although the overall chemical composition does not change much, and most crude oils and natural gases fall in a small bulk chemical composition range (Table 3-1), the diversity of

Table 3-1 Composition of Typical Petroleum

Element	Crude Oil (percent)	Natural Gas (percent)
CARBON	82.2–87.0	65–80
HYDROGEN	11.7–14.7	1–25
SULFUR	0.1– 5.5	trace–0.2
NITROGEN	0.1– 1.5	1–15
OXYGEN	0.1– 4.5	—

(From Geology of Petroleum. Second Edition. by A. I. Levorsen. W. H. Freeman and Company. Copyright © 1967.)

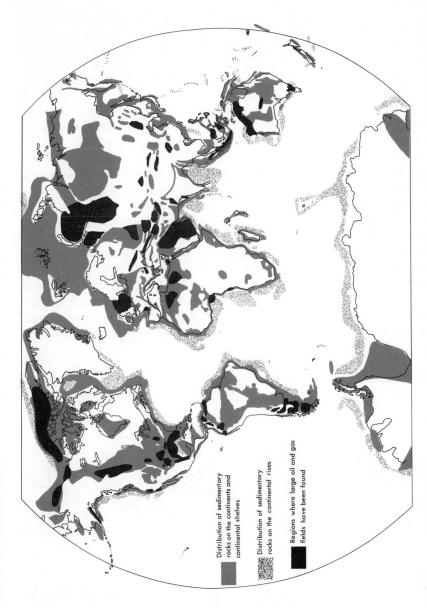

Distribution of sedimentary
rocks on the continents and
continental shelves

Distribution of sedimentary
rocks on the continental rises

Regions where large oil and gas
fields have been found

FIG. 3-5 Sedimentary rocks and regions where oil and gas have been located. The continental shelf and slope, extending out to a water depth of 2,000 meters, contain large potential resources of petroleum, particularly where they are underlain by seaward projections of sedimentary basins. Thick piles of young sediments beyond the continental slope, a region sometimes called the continental rise, are also believed to contain large petroleum resources.

individual compounds produced is so great that no two oils ever contain the same molecular mix.

The lighter and more mobile the petroleum hydrocarbons become, the more readily they migrate. Though the exact mechanisms of migration remain uncertain, oil, gas, and water in sedimentary rocks move and slowly escape toward the surface; where barriers or traps are interposed in the migration paths, accumulations result (Fig. 3-6). The fact of slow escape is substantiated by the observation that the highest ratio of oil pools to volume of sediments is found in the youngest group of oil-bearing sediments, deposited no later than 2.5 million years ago. It is also substantiated by the observation that the total amount of trapped oil decreases the farther we move back in time (Fig. 3-7).

Natural gas, which is the lighter hydrocarbon fraction, and in particular methane, may range from a small quantity dissolved in the crude oil, through a gaseous capping over an oil pool, to a separate large accumulation not associated with a nearby oil pool. All such accumulations are valuable, and the technological mastery of pipeline-laying and more recently of commercial liquefaction has made natural gas widely available as a fuel.

Petroleum, like coal, is widespread but unevenly distributed. The reasons for such a distribution are not so obvious as they are with coal. The kinds of rock that can be oil-bearing are very widespread, but the formation of oil pools is a matter of delicate timing. Most oil and gas fields are found in traps that formed by flexing and fracture of the host rocks. The traps must, of course, form before most of the oil and gas migrate away and escape, and this seems

FIG. 3-6 Types of oil traps and reservoir rocks, and the depth of known oil pools, together with the percentage of the world's oil production from each. (From *Man's Physical World* by J. E. Van Riper. Copyright 1962 by McGraw-Hill, Inc. Used with permission of McGraw-Hill Book Company.)

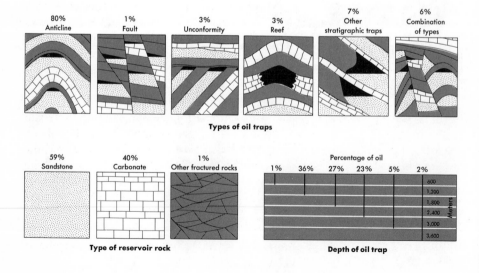

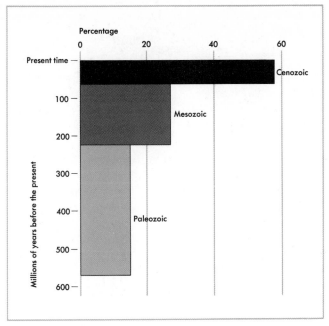

FIG. 3-7 Estimated production of crude oil from rocks of different geological age ranges. (After G. C. Gester, 1948.)

to be a rather chancy business. Geologists estimate that the chance of petroleum formation coinciding with formation of suitable large traps is so low that no more than 0.1 percent of the organic matter deposited in oil-source rocks is eventually trapped in oil pools.

Production and consumption of petroleum have reached such tremendous proportions that it is sobering to remember that commercial production began only in 1857—little more than 100 years ago—in Rumania; this was followed two years later by production in the United States. World consumption of both crude oil and natural gas is rising at a rate of approximately 7 percent a year, which corresponds to a doubling period of ten years (Fig. 3-8). So vast is this use of oil and gas that the question of supply adequacy has become vital.

Although the oil and gas industries maintain adequate reserves for near-term production, they lack a sure way of estimating the potential resources still to be discovered. Oil and gas pools are small compared with coal fields and can be located only at high cost and with considerable difficulty. However, we know where sedimentary rocks are to be found in the world, and if we assume that the oil potential in unexplored areas is just as good as it is in those most highly explored, it is possible to make a geological guess at the world's ultimate recoverable resources of oil and gas. Various authorities have done this and have come up with answers that, so far from suggesting accord, differ by factors of three or four. Part of the uncertainty lies in the percentage of oil accepted as

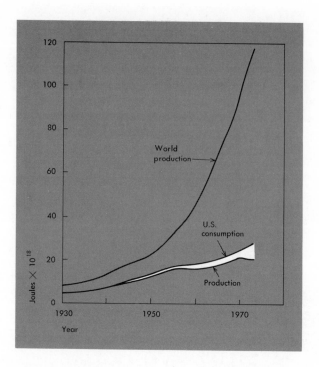

FIG. 3-8 World production of crude oil is growing at approximately 7 percent a year. U.S. production is now leveling off and at present is unable to satisfy the demand. The shaded area between the U.S.'s production and consumption curves represents the increasing quantities of oil the U.S. must import each year. Oil production is commonly reported by the barrel. A conversion of 6.0×10^9 calories per barrel was used. (Data from U.S. Bureau of Mines.)

recoverable from an already located field. The present figure of 35 percent recovery, for example, is considered too low by some, realistic by others. Another source of uncertainty lies in one's evaluation of just how well explored the sedimentary basins of a developed area like the United States actually are. Combining oil and gas (by taking 1,470 cubic meters of gas as equal in heating capacity to 1 barrel of crude oil, or 6×10^9 joules), most estimates for the world's recoverable potential resources of petroleum lie in the range 1.0×10^{22} joules to 2.5×10^{22} joules. These extremes include the oil and gas that remains to be found and developed on the continental shelf, on the continental slope, and in the great pile of sediments that have accumulated along the base of many continental slopes.

Three important conclusions can be drawn from our present knowledge of oil and gas resources. First, it seems apparent that many regions which are now major producers are also regions where future discoveries are most likely to be made (Fig. 3-9). In part this is because productive sedimentary basins on land promise to be equally so, or even more productive, in their submarine extensions. In part, it is probably due to man's ability to locate the best prospects first—

although as the recent discovery of huge gas fields in the North Sea shows, even astute prospectors like those from energy-hungry Europe can sometimes be misled. The second conclusion to be drawn from Fig. 3-9 is that oil and gas resources are like most mineral resources in that they are very unequally distributed around the world. As with coal, it seems to be the peoples of the Southern Hemisphere who have the smaller share. The third conclusion comes from the estimated size of the oil and gas resource and the world's annual consumption rate, which reached 120×10^{18} joules in 1973. Even if, on a worldwide basis, the total oil and gas resource amounts to an optimistic 2.5×10^{22} joules, the present 7 percent a year growth rate would, if maintained, consume half the total by the year 2003, and the entire amount by about the year 2015. This

FIG. 3-9   Known and probable resources of crude and natural gas, both on land and offshore, are principally in North America and Asia (including the U.S.S.R.). Bar graphs for percentage of world petroleum resources include the amount already consumed. Production patterns are changing rapidly as the North American percentage declines and as the Asiatic percentage climbs. (Data principally from U.S. Bureau of Mines and from Hubbert, 1969.)

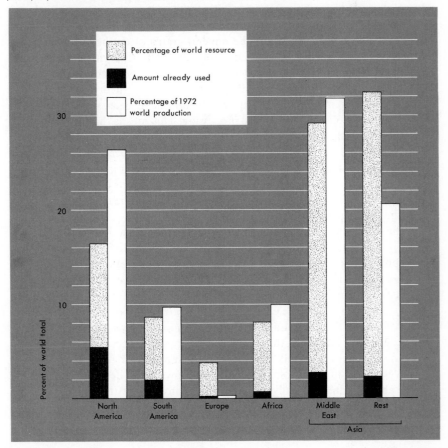

situation will not be so drastic because other fuels will be used long before this date, but the example demonstrates that the day of oil and gas in man's history may be a short one—little more than a hundred years.

Tar Sand, Shale Oil, and Other Fossil Fuels

When crude oil is pumped from an oil pool as much as 60 percent of the oil originally in the pool remains trapped in pockets between mineral grains, as coatings on minerals, and in the innumerable tiny holes and fractures of rocks. The trapped fraction of an oil pool is, in effect, not recoverable by present methods. Some success in recovering trapped oil has been achieved by blasting and fracturing the reservoir rocks, and by heating the ground to make the oil flow more easily, but secondary recovery methods still leave as much as half of the original oil in the ground. Trapped oil thus constitutes a potential resource as great as that of flowing crude oil, but it is a resource that will require very innovative technology if we are ever to see it used.

Oil which is particularly thick and viscous is called heavy oil, or, more colloquially, tar or asphalt. Heavy oil is a crude oil in that it contains large-molecule hydrocarbons. However, it does not run, flow, or migrate, but remains in place as a cementing agent for the mineral grains of the host sand. For the hydrocarbons to be recovered, the rock itself must be heated with steam to make the asphalt flow; the resulting tarry extract must then be processed to recover the valuable oil fraction. The heating process can, in theory, be performed underground, but in practice the only way tars have been worked so far is through mining, followed by surface processing. This means that for deposits to be useful, they must be shallow enough to be reached by mining. Fortunately, the largest deposit in the world, which lies in northern Alberta and is known as the Athabaska Tar Sand (Fig. 3-10), can be reached by mining. These sands are now being exploited in a small way—about 6,500 metric tons of oil are recovered each day from a 100,000 ton-per-day mining operation near Fort McMurray. Other and larger plants are planned.

The Athabaska Tar Sands cover an area of at least 75,000 square kilometers, reach thicknesses of 60 meters, and if we assume a 50 percent recovery rate, represent a reserve of 0.31×10^{22} joules. Two other very large tar sand deposits are known. The Orinoco deposit in Venezuela is estimated to be almost as large as the Canadian deposit, but it is deeper and will be more difficult to recover. In the U.S.S.R. is the Olenek deposit, reported to be about the same size as the Athabaska Sands, but again it suffers from recovery problems. No other large deposits are known in the world, but many small ones have been located during the search for crude oil. If we make the assumption that direct exploration for heavy oil will turn up some presently unknown large deposits, and that at least 50 percent can eventually be recovered, many authorities suggest that the heavy oil potential is equal to the potential crude oil plus natural gas, namely, 2.5×10^{22} joules.

A final kind of fossil organic matter is an even larger resource than coal, but unfortunately much of it is unlikely to ever be recovered. All sedimentary rock contains some organic matter, and one common variety, shale, often contains enough so that petroleum can be extracted by distillation. It is estimated that the shale resource is at least 10^{26} joules and may be larger. Processing the shale uses energy for mining and heating. It can be calculated that energy equivalent to burning 40 liters of oil would be used in processing a metric ton of shale. Most shales won't yield 40 liters of oil per metric ton, however, so unless some new and innovative way can be found to recover the oil, ordinary shale cannot be considered a potential energy resource. But shales richer than 40 liters to the metric ton can be considered. The richest deposits known, in Estonia, yield up to 320 liters to the metric ton.

The United States is fortunate in possessing the world's largest known reserves of rich oil shale. During the Eocene epoch, three large

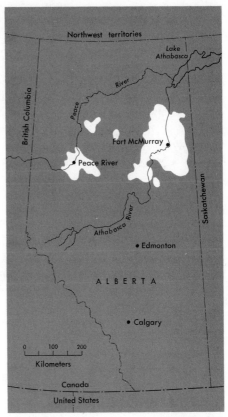

FIG. 3-10 Areas in Alberta, Canada, known to be underlain by tar sands. The first commercial production has begun near Fort McMurray.

shallow lakes existed in the intermontane area of Colorado, Wyoming, and Utah; in them was deposited the series of rich organic materials that are now oil shales (Fig. 3-11). Commonly known as the Green River Oil Shales, they are capable of producing up to 240 liters of oil per metric ton. Reserve estimates by the U.S. Geological Survey are enormous. Considering only shales capable of yielding 40 liters or more per metric ton, and assuming that half of the shales can be mined and processed, the Green River Shales are a potential resource of 0.75×10^{22} joules. Most of this potential is in the Piceance Basin of Colorado.

Production of oil from shale has been successfully carried out in Estonia, in the U.S.S.R., and in China for many years. Though minor scattered attempts at commercial production of the huge Green River Shales have been made, none has yet proven successful. Large-scale commercial exploitation in the Piceance Basin is now planned for the late 1970's, and the possible side effects from the plan have many people concerned. Mining will have to occur on a vast

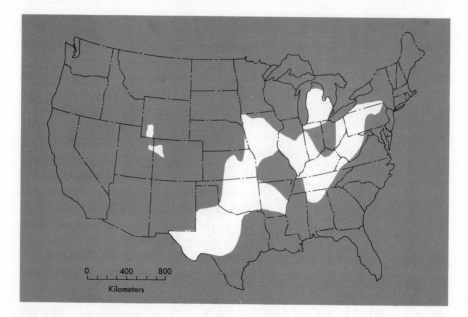

FIG. 3-11 A large part of the United States is underlain by organic-rich shales close enough to the surface to be mined and capable of yielding 40 liters or more of oil per metric ton. The richest oil shales are found in Colorado, Utah, and Wyoming. (After U.S. Geological Survey, 1969.)

scale and distillation of the shale will produce a fine, sooty residue that is puffed up and therefore larger in volume than the original shale. Disposal of this unpleasant residue is viewed as a major problem.

The United States contains many other oil shale resources in addition to the Green River Shales, but none is so rich nor so readily treated. The U.S. Geological Survey estimates that with a yield of 40 liters from sediments at least 1.5 meters thick and a mining efficiency of only 50 percent, the United States' potential resource might be as large as 7×10^{22} joules. Much of this potential lies in the central and eastern part of the country (Fig. 3-11). To mine such a huge area seems unthinkable, and once again we seem to be seeking new and innovative technology.

Rich oil shale resources in other parts of the world have not been adequately explored, but other huge deposits have already been found. The Irati Shales in southeast Brazil, perhaps half as large as the Green River Oil Shales in the United States, are the biggest units known outside North America. Lower grade resources, however, are just as abundant on other continents as they are in the United States. Accurate assessments have not been made. But if the 40 liter yield and 50 percent recovery figures are used, informed estimates by U.S. Geological Survey scientists indicate a worldwide potential resource of 10^{24} joules.

COMPARISON OF FOSSIL FUEL RESOURCES

The resource numbers derived for the fossil fuels are all very large (Table 3-2), but this cannot lead us to a feeling of complacency. The two fuels we have

Table 3-2 Potential Resources of Fossil Fuels*

Fuel	Amount in the Ground (joules $\times 10^{22}$)	Amount Possibly Recoverable (joules $\times 10^{22}$)
COAL	42	21
OIL AND GAS (flowing)	2.5	2.5
TRAPPED OIL (nonflowing)	2.5	0–?
HEAVY OIL (tar sands)	5.0	0.5–2.5
OIL SHALE		
(more than 40 liters per metric ton)	200	1.0–?
(less than 40 liters per metric ton)	10,000	?
1973 WORLD OIL AND GAS CONSUMPTION	0.012	
1973 WORLD ENERGY CONSUMPTION	0.03	

* 1 metric ton of coal is taken to be equal to 27.2×10^9 joules.
1 barrel of crude oil is taken to be equal to 6×10^9 joules.

come to rely on most heavily, oil and gas, are the two least abundant. But their production is also the least expensive in terms of manpower and disruption of the environment. Large-scale shifts to other fossil fuels will bring, inevitably, rising prices and many new problems. One very clear conclusion that can be drawn is that we would be wise to turn increasingly to sources of energy that don't involve the mining of fossil fuel. In the next chapter, therefore, other possible energy sources at the Earth's surface will be discussed.

four

energy
for the future

Energy is the sine qua non of a modern society's ability to do the things it wants to do. Such goals as maintaining the standard of living for a growing population, national security, improved quality of life, increased affluence and increased assistance to less developed societies can only be attained with increasingly large amounts of energy. While lower energy costs allow a society more freedom of action in seeking its goals, the availability of energy is the first requirement of having any freedom of action at all. (Report to the President of the United States, by Dixy Lee Ray, December, 1973.)

If supplies of fossil fuels are going to be strained, what will take their place? Dr. Ray's statement, quoted above, makes it clear how essential it is that answers be forthcoming. The world already uses energy from sources other than fossil fuels—sources such as hydroelectric schemes, nuclear power plants, tidal power plants, windmills, and geothermal steam. But as discussed in Chapter 3, society has built a massive dependence on fossil fuels and all of the other energy sources supply less than 5 percent of the total. In the future the percentage must increase because fossil fuels cannot long bear such a large burden. In this chapter, therefore, an attempt has been made to put the magnitude of other energy sources into perspective.

The Earth's Energy Flux

Before evaluating specific resources, let us consider the Earth's energy budget (Fig. 4-1). Energy reaches the Earth's surface from three sources.

The first energy source is the tides. Though tiny compared to the other sources, the tides are a renewable resource and certainly contain enough energy to merit attention.

The second source of energy comes from the Earth's interior in the form

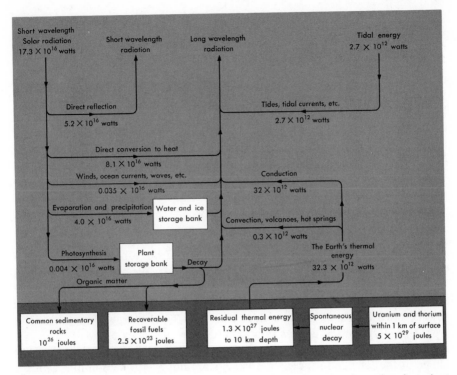

FIG. 4-1 Energy flow sheet for the surface of the Earth. Energy reaches the surface from short wavelength radiation given off by the Sun, from tides, and by heat flowing out from the Earth's interior Because the temperature of the surface is constant, the energy radiated into space must be just equal to the energy reaching the surface. (Adapted from M. K. Hubbert, 1962, Pub. 1000-D, Committee on Natural Resources, National Academy of Science.)

of heat, commonly called geothermal power. The Earth's heat is derived from disintegration of natural radioactive elements such as uranium and thorium. When the atomic disintegration process is artificially speeded up, as in nuclear power plants, the Earth's supply of uranium and thorium becomes a nonrenewable resource of vast magnitude. The nuclear energy resource shown in Fig. 4-1 considers only the uranium and thorium in the upper 1 kilometer of the continents. The nuclear resource is enormous and will never be entirely used because it is unthinkable that the entire land surface will be processed to a depth of a kilometer. But the huge nuclear number underscores an important point. *Energy is not in short supply. The questions to ask are how to use it safely and responsibly.*

Thirdly, there is the Sun, which warms the surface by heat and light. Solar energy reaches the Earth at the rate of 1.5×10^{22} joules per day, or to put it in power terms, 17.3×10^{16} watts. Some of the incoming rays are simply reflected by clouds and by the Earth's surface, but most rays pass into or through the atmosphere and follow the many paths shown in Fig. 4-1. By heating the atmosphere, the oceans, and the land, solar energy produces winds, rains, snowfalls, and ocean currents. Eventually, however, all of the energy is again radiated back into cold space and the Earth's surface remains in thermal balance. A small fraction of the solar-derived energy is temporarily stored in water reservoirs such as lakes and rivers. Another small fraction is stored in living plants and animals. Wood and hydroelectric energy from streams are therefore both renewable energy sources. Another, but much longer term storage of solar energy occurs, as discussed in Chapter 3, when organic matter from dead plants and animals is trapped and buried in sediments as they go through the long rock cycle.

ENERGY FROM TIDES

Tides are a result of the gravitational pull of the Moon and the Sun. Of the two, the effect of the Moon is by far the largest. We can picture two tidal bulges, one adjacent to the Moon, a second counterbalancing it on the opposite side of the Earth. As the Earth spins on its axis, the bulges move and everywhere produce two high and two low tides each day. Tidal energy comes, therefore, from the energy of the Earth's rotation.

The simple picture is complicated by effects from the Sun and the irregular shape of the Earth's surface. Tidal heights are not everywhere uniform. In the deep ocean they rarely exceed a meter, but over continental shelves they may reach 20 meters. Movement of such vast masses of water requires a lot of energy, and the total tidal power is estimated to be 2.7×10^{12} watts. Throughout a year this amounts to 0.85×10^{20} joules.

Utilizing the rise and fall of tides presents many problems. In France, on the River Rance, and in the U.S.S.R. on Kislaya Bay, plants already generate electricity from tidal power. These experiences demonstrate that tidal heights of 5 meters or more, and easily dammed bays or estuaries, are needed for tidal plants to operate competitively with other electrical power plants. Unfortunately, most coastal tides are about 2 meters high, and no more than 30 places around the world satisfy the right conditions. The most important are the Bay of Fundy in North America, the French coast along the English Channel, British river estuaries facing the Irish Sea, the White Sea coast in the U.S.S.R., and the Kimberly coast of northwestern Australia. If each site were fully developed, and if the schemes were so efficient that 20 percent of the local tidal energy could be recovered, the world's tidal power capacity would be 0.032×10^{12} watts—not much by comparison with our present-day needs, but very large for those few favored areas where high tides occur. Tidal power may well have an important local future, and it seems to be the one energy system with no serious environmental drawbacks.

ENERGY FROM THE SOLID EARTH

Geothermal Energy

Anyone who has been down a mine knows that rock temperatures increase with depth. Measurements made in deep drill holes from around the world show increases from 15°C to 75°C per kilometer; by projection this means that temperatures of 5,000°C or more must be reached in the core. Considering the size of the Earth, a vast amount of heat energy is stored within it.

Heat flows from a hot body to a cold one and as a consequence there is a slow outward flow of heat from the Earth. The flow averages 6.3×10^{-6} joule per square centimeter per second, or to put it in power units, 32.3×10^{12} watts across the Earth's entire surface. The total amount is vast but it is very diffuse. If all the heat escaping from a square meter could somehow be gathered and used to heat a cup of water, it would take four days and nights to bring it to a boil.

Despite the heat loss, the Earth is not cooling down; new heat is added continually. Trace amounts of several naturally radioactive atoms, principally uranium-238,* uranium-235, thorium-232, and potassium-40, occur throughout the Earth. Each time that a radioactive atom disintegrates a tiny amount of heat is released. Atoms in average igneous rock in the continental crust, for example, release 9.4×10^{-8} calorie per gram per day. This is not much, but summed over the whole Earth, it is sufficient to keep a nearly constant temperature distribution. The rate at which new heat is added is so low that we could never harness it, but the accumulated heat from millions of years can be used. But if it is used at a faster rate than it is replenished, geothermal heat must be considered a nonrenewable resource.

Assessing geothermal energy is complex because of a definition problem. What is meant by geothermal energy? If we limit our discussion to the upper 10 kilometers of the continental crust, because wells have never been drilled any deeper, and calculate all the heat energy above the 15°C average temperature of the surface, the number is enormous—1.3×10^{27} joules. But not all of this energy can be considered a potential resource. How could the energy ever be recovered? Heat now reaches the surface by conduction through rocks or, in a few special places, by outward flow of heated waters. The only ways to speed up the process are to use better conductors or to speed up flow of heated water. Replacing rock with better conductors seems unlikely. It has been suggested, however, that at some far future time we might drill deep holes, shatter rocks by controlled nuclear blasts, then circulate water which could be heated, recovered, and used to generate electricity. Schemes such as this,

---

* Uranium-238 designates an atom that contains 146 neutrons in addition to 92 protons for a total of 238 particles in its nucleus.

however, have numerous practical difficulties. Many experts do not, therefore, count the average heat in rocks as a potential resource.

Two special circumstances do offer a good deal of promise for geothermal power. The first occurs where abnormally hot rocks are found close to the surface, as in many volcanic regions. Drilling may not have to be deep and explosions may not have to be very large (or even nuclear) in order to create favorable artificial circulation systems. An experiment of this kind is now being planned for the Jemez Mountains of New Mexico, a recently extinct but still hot volcano. The second special circumstance also occurs where abnormally hot rocks occur near the surface. A rock unit which is both porous and permeable can serve as an underground water reservoir (Fig. 4-2). The hot water and steam in geothermal pools can be drawn out through drill holes and used to drive electrical turbines. When the rocks are particularly hot the fluid may be steam, when less hot a mixture of steam and water or, as in most geothermal pools, simply hot water. Because steam is preferred for electricity-generating turbines, the only geothermal pools extensively developed so far contain steam. They are used in Iceland, Japan, Italy, California, Mexico, and New Zealand, but many others are planned for future development. In a few places, geothermal hot water is used for space heating and for agricultural hot houses, but such uses are minor compared to electrical generation.

FIG. 4-2 Geothermal reservoirs occur where a water-saturated, porous, and permeable rock unit becomes heated. Heating occurs most commonly in areas of present or recent past volcanic activity, such as the western United States, Japan, Mexico, Iceland, and Italy.

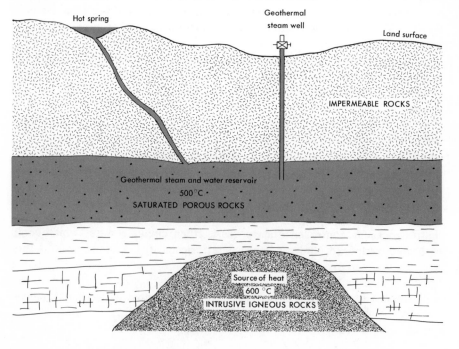

Experts from the U.S. Geological Survey report that down to a depth of 3 kilometers, which is a limit for the occurrence of big geothermal pools, the worldwide reserves of geothermal power are 8×10^{19} joules. This seems a discouragingly small amount and suggests that geothermal energy, like tidal energy, will be locally important but globally insignificant. The total heat energy in the pools is, of course, much larger than 8×10^{19} joules. But the estimate takes note of the low efficiency with which electricity can be generated from geothermal steam; experience from Iceland, New Zealand, and Italy suggests that no more than 1 percent of the energy in a pool can be effectively recovered.

Reserves of geothermal energy in natural reservoirs therefore represent only a tiny fraction of all geothermal heat. Experts cannot agree how much of the remainder should be considered a potential resource. The amount considered recoverable depends on one's optimism for future technological advances. Readers can choose for themselves. The writer tends to be skeptical that we will ever get much energy from hot, dry rocks, and therefore he puts the potential resources at 10^{20} joules, about the same level as the reserves.

Nuclear Fission Energy

Nuclear energy arises from a process first deciphered by Albert Einstein, who showed in 1905 that matter and energy are equivalent. Atomic nuclei are built up from neutrons and protons. When the mass of an atom is determined, however, it is always slightly less than the sum of the masses of individual protons and neutrons. Helium, for example, contains two protons and two neutrons and should weigh 4.03303 atomic mass units. In fact, helium only weighs 4.00260 units. The missing mass was converted to energy when the particles joined together to form the nucleus.

Disaggregation of the nucleus can only occur through the addition of enough energy to replace the missing mass. The missing mass therefore acts as a sort of glue to hold the particles together and is called the *binding energy* of the nucleus. Binding energy per proton or neutron in the nucleus varies with the total mass of the atom (Fig. 4-3). The higher an atom sits on the binding energy curve, the more energy is released when it is formed from its constituent particles.

From Figure 4-3 it can be seen that two kinds of processes will release nuclear energy. First, there is the fusion of light elements such as hydrogen and lithium to form heavier elements. This is the process that goes on in the Sun and in hydrogen bombs. Although fusion has not yet been achieved as a controlled reaction, some scientists believe that it might be a principal energy source for the future. Second, there is the disintegration or fission of heavy elements such as uranium or thorium to form two or more lighter atoms sitting higher on the binding energy curve. Fission is the process that occurs in the atom bomb. Because the reaction can be controlled, fission has already been used for the generation of electricity by nuclear power plants. We will therefore consider fission-released energy first.

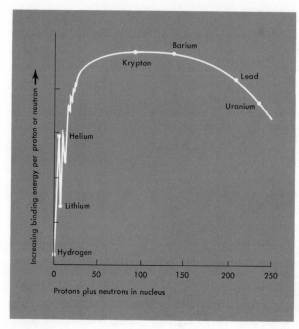

FIG. 4-3 Binding energy, produced by conversion of some of the mass of each proton and neutron in an atomic nucleus, varies with the number of protons and neutrons. The higher an atom sits on the curve, the more energy will be given off when protons and neutrons combine to form its nucleus.

Some radioactive elements can be made to fission when neutrons are fired into the nucleus, making the atoms unstable and subject to spontaneous disintegration. If additional neutrons are released by the disintegration process, a sustained, or chain, disintegration can occur. The only naturally occurring fissionable atom is uranium-235, which comprises 0.7 percent of all natural uranium atoms. The uranium-235 chain reaction was first achieved by Professor Enrico Fermi on December 2, 1942, in one of the most important experiments ever performed.

The cost of separating uranium-235 atoms from more common uranium-238 atoms is high. Once separated, however, the disintegration of a single atom releases 3.2×10^{-11} joule of energy; because a single gram of uranium-235 contains 2.56×10^{21} atoms, fission of a gram of uranium produces 8.19×10^{10} joules—equivalent to the energy released when 2.7 metric tons of coal are burned. Many uranium-235 power plants are now generating electricity. In the United States alone, 24 power plants were operating 42 nuclear reactors in mid-1974; 56 more reactors were under construction, and an additional 101 had been ordered for future installation. Around the world, by mid-1974, 77 nuclear power stations were operating and 76 more were under construction. (See Fig. 4-4.) The vital question with uranium-235 plants is one of economics. Can the uranium fuel be produced less expensively than equivalent amounts of coal or oil? The richest mines can produce uranium for about 1.9 cents a gram. Because uranium-235 is only 0.7 percent of the mass, the base cost of uranium-235 without any allowance for the difficult and expensive separation of uranium-

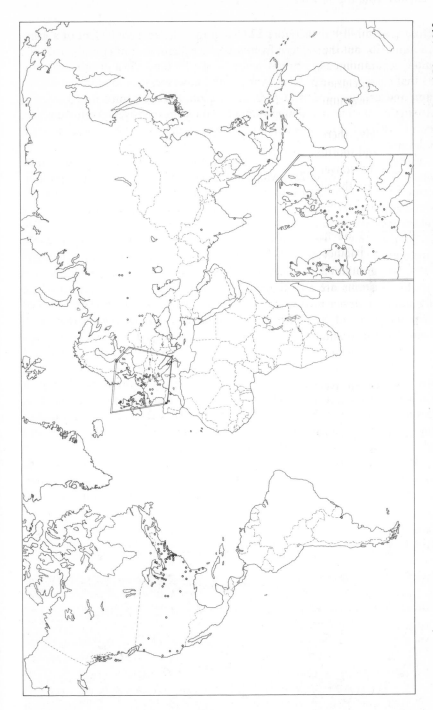

FIG. 4-4 Locations of uranium-235 power plants that were operating or under construction in mid-1974. When all the plants are completed, they will have an electrical generating capacity in excess of 50×10^{10} watts. (After International Atomic Energy Agency.)

235 from uranium-238 is therefore $2.65 a gram. The coal equivalent of a gram of uranium-235 can be produced from the least expensive mines for about $8.00. Uranium separation costs are not made public by the U.S. and other governments that carry out the process. Apparently, however, they are less than $5.35 a gram, and uranium-235 is at present an economically competitive fuel. The proviso in this conclusion is that uranium be mined inexpensively. Unfortunately, reserves of rich uranium ores that will yield uranium for 1.9 cents a gram seem to be limited.

Fortunately, there is a way out of the uranium resource dilemma. Uranium-238 and thorium-232, which are much more abundant than uranium-235, can both be converted into fissionable atoms. When a neutron with the right velocity hits the nucleus of a uranium-238 atom it is absorbed, causing two electrons to be emitted and the atom to be converted to plutonium-239. Similarly, thorium-232 can be converted to uranium-233. Both plutonium-239 and uranium-233 are fissionable and can sustain chain reactions. The trick, therefore, is to balance production of neutrons from fissioning so that the conversion of uranium-238 or thorium-232 equals or exceeds the rate at which the fissioning atoms are used up. When this happens, the process is called *breeding* and the device in which it occurs is called a *breeder reactor* (Fig. 4-5). The same amount of heat is produced from uranium-238 and thorium-232 reactions as from uranium-235. But with breeder reactors the cost of the fuel becomes less significant: At 1974 prices, the cost of raw uranium for a breeder reactor is equivalent to coal at a cent a metric ton or to oil at one fifth of a cent a barrel. Even average rocks and sea water could probably be used as sources of uranium and the fuel still be competitive with coal and oil. For this reason, research on breeder reactors is very active and many experts believe that they will be operating commercially by the year 1995.

Most of the work on atomic energy has centered on the uranium reactions, and it is uranium that has been most actively exploited and explored. Uranium occurs in two valence states, U^{+4} and U^{+6}, and the interplay of these governs

FIG. 4-5 Schematic diagram showing the difference between the uranium-235 (U^{235}) fission reaction and the uranium-238 (U^{238}) transformation into plutonium-239 (Pu^{239}) and subsequent plutonium fission. Heat produced in both kinds of reaction can be converted to electricity.

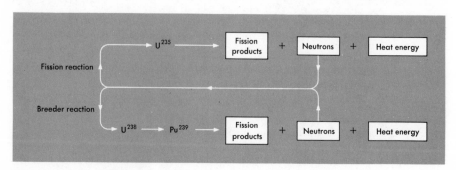

its distribution in the crust. In certain igneous rocks, uranium is widespread as the mineral uraninite (sometimes called pitchblende) which has uranium in the U^{+4} state. The U^{+4} state is readily oxidized to U^{+6} under conditions at the Earth's surface, in which form it will combine with oxygen to form the complex uranyl ion $(UO_2)^{+2}$ which itself has a positive valence and can form separate compounds. The uranyl ion forms soluble compounds, such as uranyl carbonate, which facilitate the movement of uranium in surface waters. Precipitation occurs when the solutions encounter a reducing agent, such as organic matter, that returns the uranium to the less soluble U^{+4} state. The reduced uranium may precipitate as uraninite, as it has done in many of the famous Colorado Plateau deposits where buried logs have been found almost completely replaced by uraninite; or it may precipitate as uranium-rich compounds, as it has done in many of the organic-rich shales around the world, such as the 100,000 square kilometers of Chattanooga Shale in Alabama and Kentucky, and in the Alum Shale of Sweden and Norway. Finally, the uranium may be absorbed by another mineral, and the common mineral in which this occurs is apatite, $Ca_5(PO_4)_3(OH, F)$, with U^{+4} atoms replacing some of the Ca atoms, a form in which it apparently occurs in the Florida phosphate deposits.

The measured resources of uranium in rich deposits are unfortunately not so large as we might hope. During the years 1945–1960 uranium received the most intensive mineralogical and geological scrutiny ever accorded a metal. A great many rich deposits were located and—following the assumption that further prospecting would locate yet more high-grade deposits—confident and rosy predictions were made of large, easily won potential resources. Further work suggests that this confidence was misplaced. A renewal of intense prospecting, commencing in the late 1960's, has not located the hoped-for bonanzas.

The richest deposits in the United States are apparently secondary concentrations in sedimentary rocks, arising from the movement of uranium by ground water. The largest and most numerous deposits are found in rocks of the Jurassic and Triassic periods on the Colorado Plateau of western Colorado, eastern Utah, northeastern Arizona, and northwestern New Mexico. Other rich deposits in sedimentary rocks, but from the Cenozoic, are found in Wyoming. All of these deposits, together with smaller and less valuable occurrences in numerous other states, account for a reserve of uranium reported by the International Atomic Energy Agency to be only 262,000 metric tons (Table 4-1). The reserves reported for Canada are principally located in rich vein deposits in the Great Bear Lake region, Northwest Territories, and in the Blind River district north of Lake Huron in Ontario, where extensive sedimentary rocks of Precambrian age contain disseminated uranium minerals of uncertain origin. Large South African reserves are associated with the Witwatersrand gold deposits, where trace amounts of uranium are recovered as a by-product from gold production.

Reserves of rich uranium ores in the non-Communist world are not large. Probably the figures are conservative, because an element of secrecy and caution

Table 4-1 Reserves and Resources of Uranium in the Non-Communist World*

Country	Reserves (metric tons)	Potential Resources (metric tons)
UNITED STATES	262,000	139,000
SOUTH AFRICA	200,000	62,000
CANADA	185,000	124,000
AUSTRALIA	162,000	62,000
OTHER	146,000	323,000†
WORLD TOTALS	955,000	710,000

* Reserves are considered to be deposits in which uranium can be profitably produced for less than 1.9 cents a gram. Resources are lower grade deposits that could be exploited if the price rose to about 3.0 cents a gram. (Data from *World Mining*, July 1974.)

† 270,000 metric tons are estimated for Sweden.

surrounds a strategic commodity like uranium. Regardless how much the reserves are understated, however, they are certainly not large enough to support extensive use of uranium-235 power stations very far into the future. Let us consider, for example, what would happen if all of the reserves and resources listed in Table 4-1 were used solely for their uranium-235 content. If the conversion of heat energy to electricity was 40 percent efficient, the total energy produced would be only 3.8×10^{20} joules. On the other hand, if the same efficiency factor were used for all the uranium-238 in a breeder reactor, the available energy total soars to 5.5×10^{22} joules.

Beyond the rich deposits it is difficult to make an assessment. The necessary work simply has not been done. Within the United States a general evaluation has been made of the kinds of source materials available (Fig. 4-6), and an estimate was made that if costs rose as high as $1.00 a gram, about 2×10^9 metric tons of uranium could be recovered. This enormous mass could produce 6.6×10^{25} joules of electricity. Similar figures pertain in other parts of the world. Reserve estimates for thorium are less certain than those for uranium. Because thorium is a more abundant element than uranium in the continental crust, it is possible that its potential resources exceed those of uranium. Provided that the necessary technological developments for breeder reactors can be made, therefore, nuclear fission energy appears to have a very bright future.

Nuclear Fusion Energy

Unlike fission, fusion of atomic nuclei has not yet been controlled. When achieved, fusion reactions open the way to use of nonrenewable energy resources that really are astronomic in magnitude.

The greatest amount of energy comes from fusion of the lightest element, hydrogen, to produce helium, a reaction that produces most of the energy emitted by the Sun. Similar reactions involving heavier elements produce lesser yields of energy as the masses increase, but on the other hand, they are

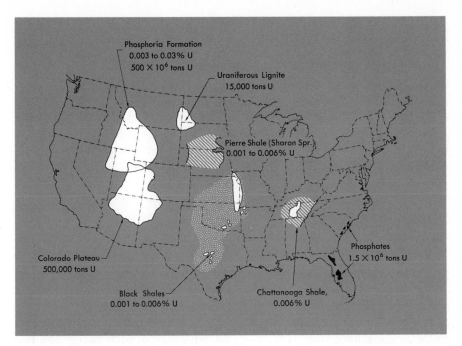

FIG. 4-6 Long-term, but low-grade, resources of uranium occur in phosphate deposits and black shales. Richer deposits, for which present reserves can be estimated, occur mostly in and around a region known as the Colorado Plateau. (After M.K. Hubbert, *Energy Resources*, Pub. 1000-D, Committee on Natural Resources, National Academy of Sciences—National Research Council, Washington, D.C., 1962.)

easier to control. The main reaction in the H-bomb, for example, involves fusion of two atoms of deuterium (a form of hydrogen with 1 neutron and 1 proton in the nucleus) to form helium. Each time this combination occurs, 7.9×10^{-13} joule is released. Each cubic centimeter of sea water contains approximately 10^{16} atoms of deuterium, and because the volume of the oceans is 1.35×10^{24} cubic centimeters, the fusion potential of deuterium in the sea is 10.7×10^{27} joules.

Hydrogen fusion, when it is attained, promises an even larger resource because there are 6,500 hydrogen atoms for every deuterium atom in the sea. With such large numbers it becomes obvious why many authorities believe extensive research into fusion reactions should be supported. Success with fusion energy, regardless of whether deuterium or hydrogen is used, is clearly not limited by fuel resources. The limitation would seem to be man's own ingenuity.

ENERGY FROM THE SUN

Solar energy has several positive features that are not combined in any other source. It is renewable, it is enormous, it is clean, and it is dependable.

The only question is how to use it. Many simple heating devices using solar energy are now in use, but most are inefficient and expensive to build. The inefficiency arises because most devices first heat water or a similar liquid and this in turn is used to drive an electrical generator. Recent research suggests, however, that devices for capturing solar heat and converting it directly to electricity with a 30 percent efficiency might be attained. With such a device, it can be calculated that about 8 square kilometers of collecting surface would be necessary for a power plant producing 10^9 watts—about the capacity of today's larger nuclear power plants. Costs of plants, the large collection areas needed, and high percentages of cloudy days in many areas will probably prevent entire countries from using solar power alone. But for the fortunate areas of cloudless skies, solar power must be considered one of the most attractive possibilities for the future.

Water Power

Solar energy need not be collected directly. Figure 4-1 shows that about 23 percent of the incoming solar radiation is used in evaporating the water which falls as rain and snow. In effect, the Sun acts as a great pump, raising water from the sea and dumping it on the land, from which it runs downhill to the sea. Running water is therefore an expression of solar energy and is a renewable resource.

Water power has been used in small ways for thousands of years, but it was only at about the beginning of the twentieth century that large-scale damming of rivers commenced for generation of electricity. Evaluation of energy resources from running water involves an assessment of the amount of water flowing in streams and rivers and how far downhill it flows before reaching the sea. Using data collected by the U.S. Geological Survey, F. L. Adams in 1962 assessed the world's potential for generating electricity by water power to be 2.9×10^{12} watts (Table 4-2). At present, less than 6 percent of this power

Table 4-2 Water-Power Capacity Around the World

Region	Power Potential (watts)	Percent Already Developed
AFRICA	7.8×10^{11}	less than 1
SOUTH AMERICA	5.8×10^{11}	1
COMMUNIST COUNTRIES	4.7×10^{11}	4
SOUTHEAST ASIA	4.6×10^{11}	less than 1
NORTH AMERICA	3.1×10^{11}	20
EUROPE	1.6×10^{11}	30
MID-EAST, FAR EAST, AUSTRALIA, NEW ZEALAND	1.1×10^{11}	20

(From F. L. Adams, after M. K. Hubbert, 1969)

has been developed, but if it were fully developed, the energy produced each year would be 0.9×10^{20} joule, about one-third of the total energy now consumed and larger than the presently installed electrical generating capacity.

There are two points of paramount importance to consider with water power. The first, which is an unfavorable point, is that even if the world's community will accept the total damming of its great river systems, the dams have definite and rather short lifetimes. A stream carries a load of suspended sediment and this material is deposited as soon as the stream is dammed. Depending on the sediment load, the reservoirs will be completely filled by sediment in periods ranging from 50 to 200 years. The great Aswan Dam recently completed on the Nile, for example, will be at least half silted up by the year 2025. Water power may be renewable, therefore, but the water power sites are nonrenewable. The second point, which is a favorable one, can be seen from Table 4-2. The world's largest undeveloped potential lies in South America and Africa. Inasmuch as these continents have very small coal resources, it is fortunate that their water power should be so plentiful. The long-term future for water power in the Southern Hemisphere must be considered very promising.

Wind and Ocean Power

If we refer again to Fig. 4-1, we see that about 46 percent of incoming solar energy is absorbed by the oceans, the land, and the atmosphere. This energy, in addition to producing winds, waves and ocean currents, warms the seas and produces all of our familiar weather patterns. At least some of it can be considered a potential resource.

Winds blow intermittently and do not readily lend themselves to large-scale power schemes. Suggestions have been made that large windmills could be erected at sea, or along the shore, for the generation of electricity. The cost of doing so, however, can be estimated and is so great that the power produced would cost several times more than power from other sources. For smaller-scale uses, as in transportation at sea, recreation, and farm use, however, wind power has long had many uses and many experts predict a resurgence of windmills and other small-scale devices in the future.

When the Sun warms the ocean by heating the surface, a thermal gradient occurs (Fig. 4-7). In theory, it is possible to draw on the ocean's vast heat reservoir by using the gradient, much as hot water is used from geothermal schemes. The difficulty is the small temperature difference involved—no more than 20°C, even in the tropics. Nevertheless, a small pilot plant using ocean thermal power was built and run for a short time by French engineers near Abidjan on the west African coast after World War II. From this experiment larger schemes have been devised and engineers have been granted patents. No one has yet convinced either private or governmental investors that the larger schemes are economical, however. If they ever do, another astronomically large source of energy will be made available. Just how large is difficult to estimate because it depends on the efficiency of the generating plant. Even if it were

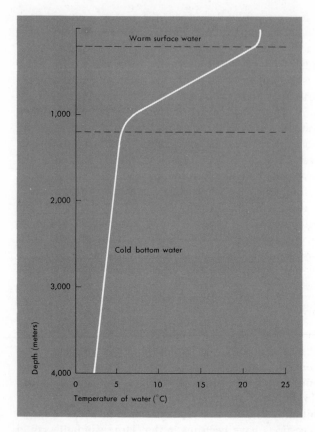

FIG. 4-7 A generalized representation of the vertical temperature distribution in the ocean. Scientists have proposed several devices for using the temperature difference between warm surface and cold bottom waters as a means of using some of the ocean's thermal energy.

less than 1 percent, the ocean's thermal energy potential exceeds the potential from fossil fuels. Perhaps it is not even necessary to use electrical generation. Scientists have already proposed and tried schemes for preserving foods such as flour and corn at the near freezing temperature of the ocean floor and for pumping cold ocean water as a tropical air conditioning system. Such uses would at least help ease the strain on fossil fuels.

Another expression of heat energy in the oceans is the great surface currents. The Gulf Stream, for example, has mechanical power from its flow equal to 2.2×10^{14} watts, or 7×10^{21} joules per year. But how such a vast renewable resource could be harnessed lies in the realm of distant speculation.

Waves are yet another expression of solar energy. Waves arise from winds blowing over the ocean, and they contain energy many thousands of times greater than in tides. For example, a single wave that is 1.8 meters high, moving in water 9 meters deep, generates approximately 10^4 watts for each meter of wave front. What vast amounts of energy are expended daily on the shorelines of the world! Wave power has been used to ring bells and blow whistles for navigational aids for many years, but only recently has large scale energy recovery been considered. The possibilities do not seem too encouraging. One patented scheme uses troughs covered with a flexible film and filled with hydraulic fluid. When the flexible trough is placed on the sea floor, the weight of a passing wave forces the hydraulic fluid through a network of pipes and into a motor, which, in turn, drives a generator.

Perhaps we should not be too skeptical about the use of solar energy resources. While the possibilities seem limited today, technologies of the future may change the whole picture. If ever a field of study beckoned people with inventive genius, it is the whole field of solar energy.

THE FUTURE

There is neither an energy crisis nor an energy shortage. There are vast amounts of energy, more than we could ever use. Any crisis is of our own making because we rely too heavily on inexpensive fossil fuel. As Chapters 3 and 4 demonstrate, there are many alternatives to oil and gas, and most offer the comforting side benefit of larger resources. But each alternative has drawbacks too. Mining of coal, oil shale, and tar sand, for example, causes serious disruptions to the land surface. Nuclear energy, the most likely energy source on the horizon, also brings problems of heroic proportions. Wastes from nuclear power plants remain lethally radioactive for periods up to thousands of years. Where can we dispose of them so that people thousands of years hence will not be threatened? We must protect generations as far in the future as the Romans lie in our past. And what is to be done with waste heat generated by large power plants? At present most water used in the cooling of nuclear power plants is returned to streams and lakes, thereby raising their temperatures. As plants grow larger, so too will the waste heat problem—unless innovative ways for its use can be developed. Other problems concern the source of capital needed to build vast power plants, the threat that exists when huge communities are dependent on plants that are vulnerable to strikes and attacks, and, of course, the inescapable problem that breeder reactors produce more and more fissionable materials that could be used for nuclear weapons.

Changing energy sources will inevitably bring changes in life styles. Fossil fuels are chemicals that can be easily transported and used in small quantities. Alternate energy sources can, in some cases, be converted to conveniently transported fuels such as hydrogen, but conversion is inefficient and expensive. Inevitably, therefore, uses such as our present systems of transportation will change.

How the story will develop is part of the intriguing future. There are only two points about this future that now seem to be certain. Energy needs will continue to rise for at least the near future, and energy sources will slowly change. As sources change so will the machines using the energy, and so too will the demand patterns for metals to build the machines. We next turn, therefore, to metallic mineral resources.

five

the abundant metals

I have for many years been impressed with the steady depletion of our iron ore supply. It is staggering to learn that our once-supposed ample supply of rich ores can hardly outlast the generation now appearing, leaving only the leaner ores for the later years of the century. It is my judgment, as a practical man accustomed to dealing with those material factors on which our national prosperity is based, that it is time to take thought for the morrow. (Andrew Carnegie, Proceedings of a Conference of Governors in the White House, Washington, D.C., May 13–15, 1908.)

Metals have versatile properties such as malleability, ductility, and high thermal and electrical conductivities. They are invaluable for technological applications, and we use the metalworking skills of a community as one measure of its development; thus most of us are familiar with such terms as the Bronze Age and the Iron Age.

Metals can be divided into two classes on the basis of their abundance in the continental crust: the *scarce metals*, with abundances less than 0.1 percent, and the *abundant metals*—iron, aluminum, manganese, magnesium, and titanium, with abundances greater than 0.1 percent.

Every rock contains detectable amounts of each abundant metal; they are usually present in separate minerals rich in the individual elements so that it is the mineral in which the element is carried that determines whether or not it is a useful resource. Abundant metals are commonly combined with silicon and oxygen, the two most abundant elements of the crust, to form silicate minerals such as albite ($NaAlSi_3O_8$), anorthite ($CaAl_2Si_2O_8$), and olivine (Fe_2SiO_4). Because silicate minerals are characteristically difficult and expensive to break down, they are undesirable sources of their contained metals. The minerals in which the abundant metals are preferentially sought are naturally those from which the metals are most easily recovered: oxides and hydroxides, such as *magnetite* (Fe_3O_4) and *gibbsite* (H_3AlO_3), and carbonates, such as *siderite* ($FeCO_3$) and *magnesite* ($MgCO_3$). Unfortunately, the desirable minerals are much less common than silicates, so the readily recovered fraction of any metal in the crust is tiny compared to the whole. Although not expressed in exactly these words, it was this very point that drew Andrew Carnegie's concern in the quotation at the beginning of this chapter.

CONSUMPTION OF THE ABUNDANT METALS

Abundant metals have such high average concentrations in the crust that small enrichments above the average leads to valuable ores. Richness of a deposit is important, therefore, but if the correct minerals are not present, it is not a sufficient factor for recovery of abundant metals. This property stands in distinct contrast to the scarce metals discussed in Chapter 6. Time has shown that Andrew Carnegie's fears can be allayed. Rich ores are certainly desirable, but provided the right minerals are present, quite lean ores of abundant metals can also be worked.

Abundant metals deserve their name on three accounts: abundance within the Earth, magnitude of man's consumption, and the rate at which consumption is increasing (Fig. 5-1). Consumption is especially high for iron, the consumption of which exceeds the aggregate of all other metals. Because supplies are plentiful

FIG. 5-1 World and U.S. production of two abundant metals. Iron is reported in the form of the ore in which it is mined because the ore is processed into different end products. The U.S., with six percent of the world's population, consumes a much larger percentage of the world production in each case, a pattern that holds true for many mineral commodities. However, the increasing divergence between the U.S. consumption curve and the world production curve reflects a rising standard of living around the world. (After U.S. Bureau of Mines.)

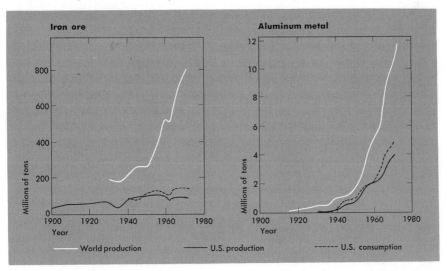

and production large, the cost of abundant metals is relatively low. For example, at the beginning of 1972, the price of pig iron was about 7.7 cents a kilogram, of aluminum 63.9 cents, and of magnesium 70.6 cents. By comparison, three scarce metals, tin, tungsten, and rhodium, respectively, cost 369 cents. 1197 cents, and 643,000 cents a kilogram. Despite the ready availability of ores, the processing of abundant metals requires advanced technology, and a lot of energy. It is not surprising, therefore, that technologically advanced countries produce and consume the largest fraction of each metal.

IRON

Iron, the second most abundant metal in the crust, is the backbone of civilization; it accounts for more than 95 percent of all metals consumed. A significant proportion of the remainder—nickel, chromium, tungsten, vanadium, cobalt, and manganese—are mined principally to be added to iron to give it more desirable properties of strength and resistance to corrosion.

Smelting iron from its oxide ores is a chemically simple process in which carbon (in the form of coke made by expelling the gaseous fraction from bituminous coal) and iron oxide react at high temperatures to form metallic iron and carbon dioxide. Because iron ore is rarely pure, limestone ($CaCO_3$) must also be added as a flux to remove impurities as slag. Selection of the best ore and production of sufficiently high temperatures are technological problems, and the history of smelting iron ores is the story of their solution. Until the beginning of the fourteenth century, all iron was produced in primitive forges by firing a charge of charcoal made from wood, iron ore, and limestone in a blast of air. These forges, a form of which was used by the people of China during the late 1960's when they boosted their iron production in backyard furnaces, were small and capable of reducing the iron oxide to metallic iron but incapable of melting the reduced iron. They instead formed a welded mass of incandescent iron grains from which slag and other impurities were removed by vigorous hammering to form *wrought iron.*

As demand for iron increased, forges were made larger, and stronger air blasts were required to fire them. As a consequence, higher temperatures were reached and iron was produced in a molten form in furnaces that were forerunners of the modern blast furnace (Fig. 5-2). By the fourteenth century *pig iron* (as the raw furnace product is called) was being produced in large quantities, and the foundations of the modern iron industry were laid. Subsequent discoveries led to: inexpensive ways to produce high-grade steels from pig iron; more efficient blast furnace procedures, including the use of coke made from coal to replace charcoal from dwindling supplies of wood; other methods to reduce iron ores; methods of handling ores containing deleterious impurities such as phosphorus and sulphur; and methods of processing low-grade ores as the richest were used up. Many sophisticated developments occurred in the nineteenth century, laying the basis for today's technology in which all types of iron ores can be successfully handled, thus assuring the availability of truly enormous reserves. These developments have made transportation costs and access to markets a greater factor in the production of iron than for that of any other metal.

Because the production of steel (and hence of the pig iron from which it is made) is large, the tonnage of additional materials is also large. Traditionally this has meant that the most profitable ores were those for which transportation costs were lowest. Britain's rise as the world's leading steel-producing country in the last century was because of the close proximity of high grade coal deposits and rich iron ores. Construction of the St. Mary's River Canal at Sault Ste. Marie in 1855 opened the inexpensive water route of the Great Lakes and brought the largest and richest iron ores then known, those of the Mesabi Range in Minnesota, into easy accessibility to the rich Pennsylvanian coking coal deposits.

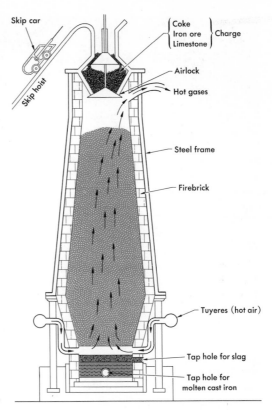

FIG. 5-2 Diagrammatic representation of the interior of a modern blast furnace. The production of each metric ton of pig iron from an ore containing 60 percent iron requires approximately 250 kilograms of limestone as a flux and a metric ton of coking coal. Recently developed electrical and oxygen furnaces require different mixes, and are noteworthy for being more efficient in their usage of coke, but the same three ingredients are used. The production of iron illustrates how interdependent are the uses of the different minerals.

Thus began the rise of the United States as a leading steel producer.

The transportation factor, though still important, is less critical now that large and efficient systems of transportation have been developed. Steel industries no longer need be set up as close as possible to the source of raw materials. Indeed, rich iron ores are now shipped by inexpensive water transportation from the far corners of the Earth to fill the growing demands of such major steel-producing countries as the United States (Fig. 5-3) and Japan, where production of rich ores is not sufficient to meet requirements. A natural consequence of this trend has been a smaller need for possession of the high-grade iron resources requisite to the industrial growth of a country, as the industrial development of iron-poor countries such as Holland and Japan attests.

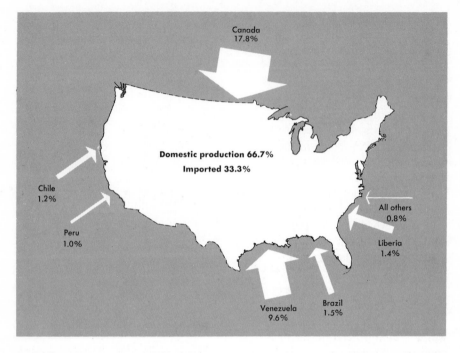

FIG. 5-3   Sources of iron ore used for production of iron and steel in the United States in 1970. (After U.S. Bureau of Mines.)

Iron Resources

Iron forms four important ore minerals (Table A-2). The property of iron that most strongly affects its concentration in nature is its ability to exist in more than one oxidation state. At the surface of the Earth, where oxygen is abundant, only the ferric or Fe^{+3} state is stable, and ferric iron minerals such as hematite and goethite occur. Beneath the surface, where supplies of free oxygen are limited, ferrous iron Fe^{+2} is stable, and ferrous minerals such as siderite, or mixed ferric-ferrous minerals such as magnetite, are found.

There are three important classes of iron deposits:

1. Deposits associated with igneous rocks
2. Residual deposits
3. Sedimentary deposits

When a magma begins to crystallize after upward intrusion into cooler portions of the crust, several mechanisms may lead to deposits associated with the resulting igneous rock. A magma is a complex liquid containing many compounds; it does not crystallize at a fixed temperature as simple liquids do, but instead it crystallizes over a temperature range, first one mineral forming, then another, until it is all solidified as an igneous rock containing several different minerals. If the early-formed minerals should be much more dense than the parent magma, they may sink rapidly and form concentrations by *magmatic segregation* on the floor of the magma chamber. When the entire magma has crystallized to an igneous rock, the segregated layers of heavy minerals may prove to be rich ores. One such is found in northern Sweden where the great Kiruna deposit occurs. Kiruna is a mass of magnetite nearly 3 kilometers long and 120 meters wide contained in Precambrian igneous rocks. Because the deposit and the rocks enclosing it have been metamorphosed, the origin of Kiruna by magmatic segregation is not entirely proven. Regardless of origin, however, this rich deposit has kept Swedish blast furnaces filled for nearly a century.

Early-formed minerals in a cooling magma tend to be anhydrous and free of volatiles such as fluorine and chlorine. As crystallization continues, therefore, the residual liquid becomes increasingly enriched, and even saturated, in volatile components. Eventually the volatiles start to escape and alter the rocks surrounding the magma chamber. *Contact metamorphic* deposits may be formed in this fashion immediately adjacent to an igneous rock. Finally, the escaping volatiles, which are commonly called *hydrothermal fluids* because they are hot and usually aqueous, may follow well-defined flow channels and, as they cool, deposit any dissolved matter they carry into *hydrothermal deposits.*

Iron deposits with igneous affiliations are usually contact metamorphic, and they are commonly small but very rich bodies of either hematite or magnetite. Only a few are large enough to be worked, although unusually large concentrations have been successfully mined in Pennsylvania at Cornwall, in Utah at Iron Springs and at Mount Magnitaya in the Ural Mountains of the U.S.S.R. Because of their relatively small size, deposits with igneous affiliations do not hold much promise of great undiscovered riches; they may be locally important but are not likely to be major factors in the future of the world's iron resources.

Residual deposits of iron minerals are formed wherever weathering occurs and the ferrous iron present in a rock is oxidized to form insoluble ferric minerals. This accounts for the brown, yellow, black, and red colors of weathered rocks and also for many of the soil colors we are familiar with. If the same weathering cycle removes more soluble minerals, the iron oxides and hydroxides remain concentrated as a residue. The process is known to have been active from Precambrian times to the present, and iron deposits formed in this way are very widespread. Commonly called brown ores, because of the color of their main mineral constituent, goethite, the residual deposits were among the first to be exploited by man. Individual deposits of rich residual ores are small, however, often being only a few tens of thousands of tons, and do not lend themselves to the large-scale mechanized mining required by the modern iron industry. The importance of rich brown ores has therefore declined in recent times and will continue to do so. Where potential resources for the future are concerned, however, residual iron deposits are of vital importance. In the tropics, where rainfall is high, soils tend to be extensively leached of soluble constituents but enriched in ferric iron compounds. These soils are commonly called *laterites*. Lateritic soils are usually too barren for concentrated agriculture, but they may contain up to 30 percent iron and although this is too low-grade to be worked today, laterites may eventually become the major source of iron. The tonnages available exceed those from all other sources by at least a factor of ten.

The sedimentary iron deposits account for most of the world's current production, as well as for reserves and potential resources for the foreseeable future. Though intensively studied for more than a hundred years, their origin is still a mystery. Today there are no known cases of the formation of iron-rich sediments similar to those found in the geological record; their origin must therefore be deduced from many different lines of evidence. The lack of modern equivalents for the ancient iron-rich sediments once again emphasizes a point stressed earlier in the book. Mineral deposits are not growing second crops.

Sedimentary iron ores are *chemical sedimentary deposits*, which means their constituents were transported in solution and deposited as chemical precipitates. This fact presents two great dilemmas in an intriguing puzzle. First, most sediments contain material transported in suspension. The iron formations are free of suspended matter which means they must have formed in unusually clear bodies of water. Second, the common iron mineral in the sediments, hematite, has the same property as other ferric iron minerals; it is essentially insoluble in sea, lake or river water. But as we have seen, weathering at the Earth's surface converts all the ferrous iron in a rock to the ferric state. How, then, could the iron have been transported in solution? The only way iron can be readily moved in the hydrosphere is to keep it in the more soluble ferrous state, which is almost impossible today because of oxygen in the atmosphere, or else somehow to change the normally neutral or slightly alkaline surface waters into acid waters, in which ferric iron is a little more soluble.

The period in the Earth's history when the greatest of the iron-rich sediments were laid down extended from 3.2 to 2.4 billion years ago. Called Lake Superior type ores after the area in North America where they were first studied, these ancient sediments are found on all continents and we can only guess how they formed. The conditions that best explain the deposits include long periods of erosion and denudation of the continental masses and the occurrence of shallow inundations by the sea. The extensive erosion that preceded inundation left little detrital material to be carried in suspension; thus, slowly accumulating chemical precipitates were relatively more important in the new marine basins. Although we have no direct proof, it is believed that at this early stage in the Earth's history the atmosphere might have had a different composition. There was probably very little free oxygen present, and possibly the content of carbon dioxide may have been higher, perhaps even as much as 100 times higher than at present. Under such conditions the surface waters would be less oxidizing and slightly acid, allowing more iron to be moved in solution and eventually to be precipitated as iron oxide and hydroxide minerals in the shallow seas.

Transportation of iron in surface waters is only one of the interesting problems associated with the Lake Superior type iron ores. The sediments are laid down in repetitive fine bands—often in layers less than a millimeter thick— of iron-rich and silica-rich layers. So striking is the banded texture that the rocks are commonly called *banded iron formations* (Fig. 5-4(A)). The silica-rich layers are now called cherts—exceedingly fine-grained quartz resulting from the recrystallization of colloidal silica—and preserved within the cherts are the fossil remains of primitive microscopic plants that were apparently growing in the shallow seas of the time (Fig. 5-4(B)). These microscopic plants, first found in the Gunflint Formation but now identified in many banded iron formations around the world, are the oldest fossils known, predating the abundant fossil record beginning 600 million years ago, in Cambrian times, by at least 2,500 million years. Whatever the still unkown circumstances for the precipitation of banded iron formations, we are forced to conclude that they must not have been unusual, for the formations are found wherever rocks of suitable age occur (Fig. 5-5); we must further conclude that conditions suitable for their formation have not been present on the Earth for at least 1.7 billion years.

Although the great banded iron formations were formed in Precambrian times, many important iron-rich sediments are known in the post-Cambrian time, a 600-million-year period in which the continuity and evolution of animal and plant life provide good evidence that the atmosphere remained nearly constant and similar to that of today. The post-Cambrian deposits are quite different from the Precambrian, however. They are usually mixed with abundant

FIG. 5-4 (A) Photograph of a specimen of banded iron formation from the Mesabi Range, Minnesota. The chert-rich bands (light) which alternate with the iron-rich bands (dark) are believed by some to represent seasonal changes due to explosive growth of silica-secreting organisms during warm summer months. (B) Fragment of a primitive microscopic plant approximately 2 billion years old from the Gunflint Iron Formation, Ontario, magnified 2,000 times. (Courtesy E. Barghoorn. From Barghoorn and Tyler, *Science*, v. 147. Copyright 1965 by The American Association for the Advancement of Science.)

A   B

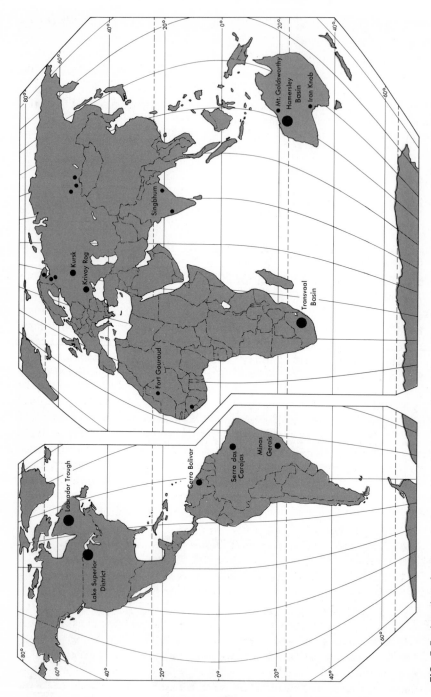

FIG. 5-5 Location of some major banded iron formations of the Lake Superior type. These formations assure abundant iron resources far into the future.

Labrador Trough

Lake Superior District

Cerro Bolivar

Serra dos Carajas

Minas Gerais

Kursk

Krivoy Rog

Singbhum

Fort Gouraud

Transvaal Basin

Mt. Goldsworthy

Hamersley Basin

Iron Knob

material transported in suspension; although the iron minerals themselves are clearly chemical precipitates, they are also limited in a real extent. The Precambrian iron sediments can be traced for hundreds and sometimes thousands of kilometers and apparently formed over a large part of the shallow sea floor. The post-Cambian deposits formed only in restricted basins, often less than 70 kilometers across, and show clear evidence of being caused by special and local, rather than worldwide, conditions: First, a warm and humid climate which allowed a thick cover of deeply rooted plants to develop on the land surface, and second, soils and permeable rocks penetrated by ground waters rich in carbon dioxide from the subsurface production by root processes, and rich in organic acids from plant decay. The resulting subsurface waters were quite acid and leached iron from the soil and rocks. The leached iron moved out via subsurface routes to the sea. When the debouching area was the open ocean, the iron was quickly dispersed, but in special cases where the debouching area was a shallow, restricted marine basin, the iron was trapped and accumulated as a chemical sediment.

Iron ores of the post-Cambrian type are important because they supply most of the iron mined in Great Britain, France, Germany, and Belgium, as well as an important fraction of the Newfoundland and Birmingham (Alabama) ores. Their worldwide importance is decreasing, however, as the European deposits are depleted and an ever greater reliance is placed on the Lake Superior type ores.

The Precambrian banded iron formations contain from 15 to 40 percent Fe, traditionally considered too low to warrant recovery. However, where the iron formations have been elevated and exposed at the surface, chemical weathering has often removed the associated siliceous or carbonate minerals and left a secondarily enriched residual ore containing 55 percent Fe or more (Fig. 5-6(A) and (B)). The great Precambrian iron deposits of the Lake Superior region and the Labrador Trough in North America, of Cerro Bolivar in Venezuela and Minas Gerais in Brazil, of Krivoi Rog in the Ukraine, and in many other parts of the world are all extensive iron formations with local zones enriched by the leaching of silica. Until recent years the unleached and therefore unconcentrated iron formations below the rich ores had not been mined, but with the depletion of the rich ore in the United States, ways have been developed to mine and beneficiate the unleached ores—called *taconites*—in the Lake Superior district. Beneficiation is a commonly used process in the mining industry to free the ore minerals from those of the enclosing waste rock and to effect an inexpensive concentration.

The mining of taconites is a good example of technological advances influencing potential resources. When first considered after World War II, taconites were thought of as expensive—even as desperate—alternatives to dwindling supplies of the rich ores in the Lake Superior district. When production of the lean ores was begun, however, it was discovered they could be beneficiated by using differences in magnetism and density between the iron

A

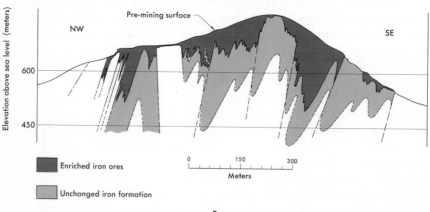

Enriched iron ores

Unchanged iron formation

B

FIG. 5-6 (A) Aerial view of Cerro Bolivar, Venezuela, and the mining operations on its summit. The hard and resistant mass of banded iron formation stands above the surrounding plain as an erosion remnant. (Courtesy of Orinoco Mining Company.) (B) A cross section through Cerro Bolivar, showing the rich iron ores formed by secondary enrichment of the original iron formation. (After J. C. Ruckmick, 1963, *Economic Geology*, v. 58, p. 222.)

minerals and the admixed silica minerals. Furthermore, it was found that (1) pellets formed from the concentrate made a stronger and more efficient blast furnace feed than traditional ores and (2) consequent savings in smelting more than offset the extra costs of mining and beneficiation. Taconite pellets have now become the standard of quality in the industry; by 1973 they accounted for more than half of the iron produced in the United States. The trend will certainly continue, and it has been estimated that by 1978 it will account for more than 75 percent of U.S. production.

Reserves of leached and enriched iron ores are large—many billions of tons—but they are miniscule compared to the amount present in the unaltered iron formations. Estimates by the U.S. Geological Survey in 1965 showed that taconite reserves in the Lake Superior region alone exceeded 10^{11} tons of iron. Other estimates revealed that even larger potential resources—in excess of 10^{12} tons at each place—were available in the iron formations of the Transvaal, the Ukraine, the Hamersley Basin of Western Australia, the Labrador Trough of northeastern Canada, and Minas Gerais in Brazil.

Considering that man's present consumption of iron is less than 10^9 tons a year, the potential resources of iron are so greatly in excess of projected needs, even assuming liberal growth rates, that it seems safe to assert that many centuries will pass before depletion of present ores becomes a serious problem. Mr. Carnegie's pessimistic remarks quoted at the beginning of the chapter were incorrectly applied to iron. They could better have been applied to bauxite, the common ore of aluminum, as will become apparent in the next section.

ALUMINUM

Aluminum is even more abundant in the Earth's crust than iron and, because of its lightness, metallic aluminum has a more desirable weight/strength ratio than iron. Although the element was first separated in a pure form in 1827, it was not until the end of the nineteenth century and the early part of the twentieth that methods capable of producing aluminum metal of high purity were developed. From that time on, the production and number of uses of aluminum have steadily increased to the point where the world's production exceeded 11 million metric tons in 1974. Aluminum has, in addition to its desirable strength and weight properties, a high resistance to corrosion and is a good conductor of electricity. Most of the technological uses take advantage of one or more of these properties, and, as a result, aluminum has challenged other metals such as iron for some of its structural purposes and copper for some of its electrical uses; it is estimated, for example, that 90 percent of the new electrical transmission lines in the United States contain aluminum conductors. Aluminum has its own distinctive properties, however, and has proved highly versatile in new uses in the construction, transportation, and packaging industries.

The production of aluminum requires large expenditures of electrical power; each ton of aluminum produced uses energy equivalent to that produced from the burning of 7 tons of coal. The aluminum industry consumes approximately 3 percent of the electric power generated in the United States. As production grows, it becomes increasingly desirable to ship aluminum ores to sources of inexpensive electrical power, such as the hydroelectric plants built along the Columbia River in the Pacific Northwest, in the sparsely populated regions of northern Canada, and in the South Island of New Zealand. High power consumption has been the most important feature in determining the distribution of aluminum producers around the world.

Aluminum Resources

Aluminum is like iron in that it occurs in many minerals. To the present, however, almost all aluminum has been produced from the three hydroxides gibbsite, boehmite, and diaspore. This use pattern introduces a curious anomaly. Aluminum is the third most abundant element in the crust, but its hydroxide compounds are among the rarer minerals. As a consequence, the reserves and potential resources of traditional aluminum ores are severely limited. Not only are the resources of aluminum hydroxide minerals severely limited, they are also severely restricted on a geographic basis.

Minerals formed by igneous processes and by metamorphism are stable deep in the Earth's crust, an environment in which large bodies of free water do not exist; such minerals are often anhydrous, or at best contain very little water. When brought to the surface, they are no longer stable; although the rate of chemical change to a more stable form is slow, they are gradually transformed into new minerals, most of which are hydrous. This transformation at the Earth's surface is called *chemical weathering*. During chemical weathering, elements such as Na, K, Ca, and Mg form relatively soluble compounds and are soon removed. The residue is left as a lateritic capping (Fig. 5-7). Most laterites are iron-rich, but some are aluminum-rich because they contain the aluminum hydroxide minerals and are called *bauxites* (after the little village of Les Baux in southern France, where they were first recognized in 1821).

Bauxites apparently form only in residual deposits caused by weathering under tropical conditions. On a low-lying or relatively flat surface, rainwater runoff is slow, and mechanical removal of weathering products in suspension is minimal. Even though the solubility of most minerals is extremely low, the most effective way of moving material in low-lying areas is in solution. The high rainfall and warm climates of tropical areas speeds the process greatly.

In addition to rainfall, topography, and temperature, the acidity of the leaching waters is important in the formation of bauxites. After the removal of most elements, a residue rich in clays, such as kaolinite $(Al_2Si_2O_5(OH)_4)$ remains. Kaolinite, too, will dissolve. If the water percolating down is very acid, it goes into solution completely; if the water is very slightly acid, only the silica

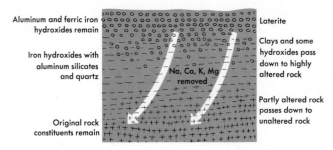

Aluminum and ferric iron hydroxides remain — Laterite

Iron hydroxides with aluminum silicates and quartz — Clays and some hydroxides pass down to highly altered rock

Na, Ca, K, Mg removed

Partly altered rock passes down to unaltered rock

Original rock constituents remain

FIG. 5-7 Leaching of the most soluble components during chemical weathering leaves a lateritic capping in which the less soluble aluminum and iron hydroxides are concentrated. When aluminum hydroxides predominate, the laterite is called a bauxite.

component of the kaolinite goes into solution, and aluminum hydroxides are left behind as bauxite.

The source rocks for bauxites have a wide range of compositions, some of which do not contain much aluminum to begin with, but most tend to have relatively low silica contents compared with other rocks. In fact, several of the world's large bauxite deposits are developed from limestones, which are essentially calcium carbonate, but which usually contain small amounts of clays and even smaller amounts of iron minerals, so that the laterites formed from them are bauxitic. In a tropical climate, limestone dissolves rapidly and the clay residues remain; the acidity of the waters leaching the limestone is apparently just right for the development of bauxite.

Bauxites are widespread in the world but are concentrated in the tropics. Even where they are found in presently temperate conditions such as southern France and Arkansas, it is clear that when they did form the climate was tropical. Because bauxites are superficial deposits, they are exposed and vulnerable to mechanical weathering processes if climatic or geologic conditions should change in any way. Bauxites are unknown in glaciated regions, for example, because the overriding glaciers scrape all the soft materials off the surface. Because of their vulnerability to later erosive processes, most bauxites are geologically young: more than 90 percent of them formed no earlier than the Cretaceous period, and the largest of all formed in the tropics during the last 25 million years.

Tropical regions are among the least developed parts of the world, and it is only since World War II that they have been extensively explored for bauxites. The results have been spectacularly successful, with the discovery of vast reserves of rich ores in the tropical regions of Australia, the Caribbean, Africa and South America where very large deposits have been recently found in the Amazon Basin.

The world's resources of bauxite were estimated by the U.S. Geological Survey in 1967 to consist of 5.8×10^9 metric tons of reserves and 9.6×10^9 metric tons of potential resources that were too lean or remotely situated to be called reserves. Approximately 73 percent of the total reserves plus resources lay in the tropical regions of northern Queensland in Australia, Guinea and Cameroon in Africa, Surinam and Guyana in South America, and in Jamaica. The remainder was widely dispersed, with the larger portions being in Asia and Europe. Since the 1967 estimates were made, new deposits have been found in South America and Africa. In addition, some of the potential resources have become reserves as technology has learned to process poorer materials and as transportation has become more efficient. The same scientist who was responsible for the 1967 estimate now suggests that if information were available, an updated 1974 total for bauxite resources would be about 20×10^9 metric tons, and at least 10×10^9 metric tons would be reserves.

The total, 20×10^9 metric tons, sounds very large but in reality it is tiny compared to the huge resources of iron ore, for which individual deposits far exceed the entire bauxite resources. Considering that even a rich bauxite will only yield about a third of its weight in aluminum, and considering that production of new aluminum exceeded 11×10^6 metric tons in 1973, the expressed concern for long-term adequacy of bauxite resources is well-based. Bauxite reserves are like oil reserves—they are adequate for the immediate future and probably reliable to the end of the present century, but they are incapable of sustaining our use patterns far into the future. Bauxite's mode of formation makes it unlikely that resources will rise by factors of ten, so the present estimate of 20×10^9 metric tons is probably realistic.

Although world resources of bauxite are adequate to last into the next century, economic and political factors make inevitable the development of recovery of aluminum from other sources. Aluminum-consuming countries are not large bauxite producers so one factor is simply the rising cost of bauxite as producing countries demand larger returns from importing countries by raising taxes, by increasing costs for land and mining rights, and by requiring large capital investments to provide jobs and bolster local economies. Other factors include rising costs of transportation, plant, and mining facilities, threats of nationalization before investments are returned and profits secured and the stability of governments. Financial pressures are so strong that experts believe some of todays reserves may again slip back to become potential resources.

There are, therefore, two pressures facing bauxite. One we might call a sociopolitical pressure, the other a pressure of abundance. Both lead to the same conclusions—alternative sources will be developed to replace bauxite. If bauxite won't serve for the future, what alternatives are there? The possible candidates for alternate supply sources are confined to the material listed in Table 5-1. At the present time aluminum is being extracted from nepheline in the U.S.S.R., from alunite in Japan, the U.S.S.R., Mexico and in the U.S.A. on a test basis, but, unfortunately, both nepheline and alunite suffer the same problems of geographic restriction and overall abundance that plague bauxite. Much more hopeful are the possibilities of producing aluminum from other sources.

Table 5-1   Current and Potentially Important Ore Minerals of Aluminum*

PRESENT-DAY ORES

Mineral	Composition	Content of Al (percent)	Remarks
BOEHMITE, DIASPORE	$HAlO_2$	45.0 ⎱	Principal constituents
GIBBSITE	H_3AlO_3	34.6 ⎰	of bauxite

ORES FOR THE FUTURE

Mineral	Composition	Content of Al (percent)	Remarks
ANDALUSITE, KYANITE, SILLIMANITE	Al_2SiO_5	33.5	Used by Sweden during World War II
KAOLINITE (the most aluminous clay; any clay-rich sediment can be used)	$Al_2Si_2O_5(OH)_4$	20.9	Successful test plants by the U.S.A., the U.S.S.R., Poland. Used by Japan and Germany during World War II.
ANORTHITE (the most aluminous feldspar)	$CaAl_2Si_2O_8$	19.4	Successful test plants in the U.S.A. and Norway
NEPHELINE	$NaAlSiO_4$	18.4	Used on limited scale in the U.S.S.R.
ALUNITE	$K_2Al_6(OH)_{12}(SO_4)_4$	19.6	Successfully tested by the U.S.A. and Japan. Used on limited scale in the U.S.S.R.
DAWSONITE	$NaAl(OH)_2CO_3$	18.7	Will be produced as a by-product of oil shale production in the U.S.A.
WAVELLITE (one of several similar minerals that occur together)	$Al_3(PO_4)_2(OH)_3 \cdot 5H_2O$	19.6	A by-product from phosphate mining. Now wasted.

* Further discussion of potential materials may be found in Bulletin 1228 of the U.S. Geological Survey, 1967.

Clays, such as kaolinite, are formed during the weathering cycle of most rocks and, following quartz, are the most abundant minerals in newly deposited sediments. If production costs could be lowered, clays could be used right now; successful test plants have already been operated in the U.S.A., the U.S.S.R., and Poland but the cost is greater than aluminum from bauxite because even more energy is needed in the reduction process. If the necessary economic step could be achieved, aluminum could be freed from resource supply problems. All countries have clays in abundance and the potential resources dwarf those of iron.

Until proven economically successful, however, aluminum from clay is only one of several possibilities. Referring to Table 5-1, we see that other opportunities exist. The aluminum silicate minerals andalusite, sillimanite, and kyanite form by metamorphism of clay-rich sedimentary rocks. Like their parent materials they are abundant and widespread, but they must be beneficiated before processing. A plant to produce aluminum from the aluminum silicates is now under construction in southern France.

Anorthite feldspar occurs in essentially pure masses and production of aluminum has been shown possible by test plants in the U.S.A. and Norway. Abundance of anorthite is hardly a limitation. To depth of 30 meters in the Laramie Range, Wyoming, anorthite deposits contain an estimated 30×10^9 metric tons, and for the United States as a whole, the figure is in excess of 10^{11} metric tons. Research and test production of aluminum from anorthite is proceeding and some authorities believe that it is a more likely successor to bauxite than are the clay minerals.

The remaining alternate sources in Table 5-1 offer fewer technological problems than clays or anorthite, but, like alunite and nepheline, seem to suffer from supply restrictions. Perhaps the most realistic assessment of aluminum is that it is a commodity with vast potential resources waiting for some innovative technological advances.

MANGANESE

Manganese is an essential additive for steels to combat small amounts of oxygen and sulfur. Up to 7 kilograms are required for every ton of steel produced, and no satisfactory substitute has ever been found.

Manganese, like iron, has more than one oxidation state, and its distribution is controlled by this property. Manganese is readily concentrated by sedimentary processes; in fact all the important manganese resources of the world are in sedimentary rocks or residual deposits formed by leaching them. As with iron, the most oxidized compounds of manganese are the least soluble and are those concentrated both in residual and sedimentary deposits.

The world's largest reserves, in which pyrolusite (MnO_2) occurs in chemical sedimentary deposits, are found in the Soviet Union, at Chiaturi in Georgia and at Nikopol in the Ukraine. These two deposits contain reserves of hundreds of millions of tons. Important residual deposits, mainly of pyrolusite and psilomelane

($Mn_2O_3 \cdot 2H_2O$), are found in South Africa, India, Australia, Brazil, Gabon, and China. Like bauxite, the rich residual deposits of manganese oxides are concentrated in the tropics where high rainfall and deep weathering are common. Though widely distributed, the world's reserve of manganese in ores amenable to mining was estimated by the U.S. Bureau of Mines in 1970 to be only about 10^9 metric tons. Compared with a present total annual production of about 20×10^6 metric tons, this is not a large figure. Possibly further exploration in tropical regions will uncover large sources of rich manganese ores, but it is likely that we will soon have to turn to new and unconventional sources.

One of the largest potential resources of manganese, and one that has excited much discussion and speculation in recent years, rests on the ocean floors. During the years 1873–1876 the Royal Society of London sent the ship "Challenger" on an epoch-making voyage around the world to gather data on the waters, animals, plants, and bottom deposits of the oceans. One of the most interesting discoveries made was of an abundance of black nodules, as much as several centimeters in diameter, on the floors of the three major oceans (Fig. 5-8). The nodules are a mixture of manganese and iron oxides and hydroxides. Although popularly called manganese nodules, a more correct term is ferromanganese nodules. Although the nodules have been intensively studied in recent years, their origin remains something of an enigma. The manganese is probably derived by normal erosive processes on the land surface and then, through a complex series of steps, slowly migrates to the ocean bottom where it accumulates. The rate at which the nodules grow appears to be extremely slow—as slow as 0.01 millimeters per 1,000 years. They are widely spread on the sea floor, however, and it has been estimated that approximately 2×10^{12} metric tons of nodules could be dredged from the bottom of the Pacific Ocean alone. These would yield 4×10^{11} metric tons of manganese.

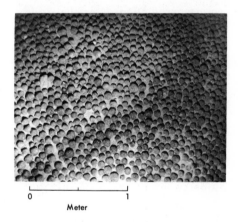

FIG. 5-8 Photograph of the deep Pacific Ocean floor, 5,320 meters deep, showing a field of ferromanganese nodules. (Courtesy of Lamont-Doherty Geophysical Observatory of Columbia University.)

Manganese nodules are already being dredged from the floor of the Pacific on a trial basis for commercial production by both Japanese and U.S. companies. It is estimated that regular production may commence as early as 1976. Curiously, this early production will not be for the manganese content, which may be no more than 25 percent, but instead for other metals. Nodules in certain areas contain several percent of copper, nickel and other scarce metals, and if separation techniques are successful, the first nodules will probably be produced for these reasons. Manganese, if recovered at all, will be a by-product. What bottom dredging of the ocean will do to sea life, who really has the right to recover the material, and how mining is to be monitored and policed are vast problems for the future. An entire new field of legal expertise will apparently have to develop.

TITANIUM

The metal titanium, like aluminum, combines light weight with high strength and high resistance to corrosion. For some purposes, for example, in the development of supersonic planes, it has answered many technological needs. It is a difficult metal to work, however, and an even more difficult metal to recover from its ores. Indeed, commercial production of titanium began only after World War II, and U.S. consumption had only reached 12,000 metric tons by 1971. It is still too early in the history of titanium to estimate accurately its ultimate importance to mankind.

By far the largest use of titanium, accounting for more than 90 percent of world consumption, is as the oxide (TiO_2), which is widely used as a white pigment for paints.

Ilmenite ($FeTiO_3$), the main ore mineral, is concentrated by magmatic segregation. When a body of magma is large, and the cooling rate slow, minerals, such as olivine, pyroxene, and ilmenite that are dense and that start crystallizing early in the cooling cycle, will settle and form distinct layers (Fig. 5-9). The resulting igneous rocks are referred to as *layered intrusions* and are important sources not only of titanium, but also of several scarce metals such as chromium, platinum, cobalt, copper, and nickel.

The largest known titanium deposits in the world, at Allard Lake, Quebec, are magmatic segregations as are the deposits at Sanford Lake in the Adirondack Mountains of upper New York State and at Blaafjeldite in Norway.

A second titanium mineral, rutile (TiO_2), is as common in the crust as ilmenite, but there are no igneous or metamorphic processes by which it is concentrated. There is, however, an entirely different way by which both rutile and ilmenite become concentrated. Both minerals are widely distributed in small amounts in igneous and metamorphic rocks; both are heavy and extremely resistant to attack from chemical weathering. Resistance to breakdown means that they are among the last minerals to disappear in the weathering cycle; when clays and other fine-grained breakdown products are carried away in suspension, the heavy, chemically resistant minerals remain behind and concentrate in *placers* (see Chapter 6). Important placer deposits are worked in Australia, Sri Lanka (formerly Ceylon), India, Sierra Leone, and New Jersey in the United States. Large reserves exist and even larger potential resources are

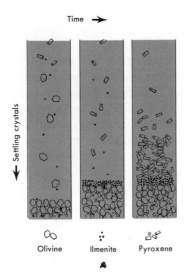

FIG. 5-9 (A) Large and heavy minerals settle faster than small and light ones during magmatic segregation, giving rise to marked compositional layering in certain igneous rocks. Ilmenite, magnetite, and chromite, the principal ore of chromium, are all concentrated in this fashion. (B) Example of layering developed between olivine and pyroxene in a layered intrusive on Duke Island, southeastern Alaska. The relief is caused by differing resistance to weathering of the two minerals. Scale is 15 centimeters long. (From H. P. Taylor, Jr., and J. A. Noble, *Economic Geology*, Monograph 4, 1969.)

known; detrital ilmenite deposits on the Atlantic coastal shelf of the United States, for example, exceed 10^9 metric tons. Reserves and potential resources of the titanium minerals seem more than adequate for the far future.

MAGNESIUM

Magnesium is the lightest metal and, being strong, is in demand for production of light corrosion-resistant alloys. The annual production of magnesium as a metal, however, is small compared with that of iron and aluminum—about 250,000 metric tons for the world in 1972. The main use of magnesium is in compounds, particularly the oxide (MgO), which has desirable thermal and electrical insulating properties. The major sources of magnesium are the sea, which contains an inexhaustible supply (see Chapter 2), and the minerals dolomite, $CaMg(CO_3)_2$, and magnesite, $MgCO_3$, both of which are widespread constituents of the crust. Dolomite occurs in marine sedimentary rocks and commonly forms an essential constituent of a rock called *dolostone*. Magnesite is found in sediments, residual concentrations, and hydrothermal deposits.

Reserves of magnesium are almost limitless and are so widely available to all nations that any discussion of their magnitude has little point. The only limits and constraints on man's use of magnesium will be his own technology.

ABUNDANT METALS IN THE FUTURE

From the foregoing discussion it seems apparent that geochemical abundance in the Earth's crust carries with it an assurance of resource abundance for man. The abundant metals do not pose problems of future supplies, but there probably are difficulties ahead as traditional ores are changed or moved to lower grades. With the exception of aluminum, none of the challenges seems likely to produce major changes in present economic structures, and even for aluminum, technological solutions will probably avoid any serious challenges to societal use patterns. With the geochemically scarce metals, to be discussed in the next chapter, a very different set of conclusions is forced on us.

six

the scarce metals

The foreseeable exhaustion of ores of some metals and the continually decreasing grade of most ore deposits now used warn that ample lead time will be needed for technology to work out such answers as it can and, similarly, to allow the economy and population density to make the necessary adjustments to changing mineral supplies. (T. S. Lovering, "Mineral Resources from the Land," in Resources and Man, *edited by P. Cloud, 1969, U.S. National Academy of Sciences.)*

The geochemically scarce metals are defined as those with crustal abundances below 0.1 percent. It is surprising to find that such common commodities as copper, lead, zinc, and nickel, all of which have large and growing rates of production (Fig. 6-1), are geochemically rare and belong in the scarce category with gold, silver, and platinum. Most experts believe that it is in this group of resources that shortages are likely to develop first, posing a serious challenge to technological development. Indeed, some shortages are already upon us. Production of silver, gold, and mercury cannot even meet present demands, let alone those of the future.

The scarce metals are a vital group that speeded development of the technological marvels of the last hundred years, such as the generation and distribution of electricity; telegraphic, radio, and television communication; aeronautics; rocketry; and nuclear power. The electrical industry, for example, would certainly have progressed more slowly without abundant and inexpensive supplies of copper. The threat posed by scarce metals for the future is that these same technological marvels, and others to come, may be imperiled. Unlimited supplies of uranium, for example, will be of little use if the essential metals to construct nuclear power plants and to transmit and use the electricity so generated are not available.

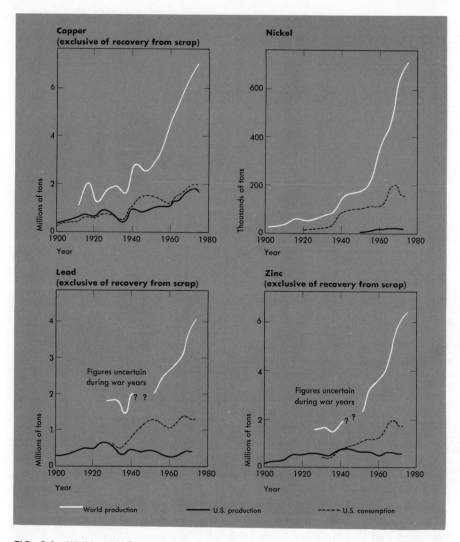

FIG. 6-1 World and U.S. production of several geochemically scarce metals, together with the U.S. consumption rate. The percentage consumed by the U.S. is steadily declining as other countries enjoy increased living standards. (After U.S. Bureau of Mines.)

The relative abundances of scarce metals in the crust are low, but the total amounts are large simply because the crust itself is large. Although some have endorsed the idea, it makes little sense to consider "average" rock as a potential source of scarce metals; the huge tonnages to be processed and the tremendous energy consumption needed to conduct the operation are prohibitive of profit. Furthermore, it would only make sense to process average rock if all the contained metals were recovered. This would lead to an oversupply of iron, aluminum, and other abundant metals and an insufficient supply of scarce metals such as tin, silver, and platinum. The reason is obvious. We do not use metals in proportion to their natural abundances. We consume scarce metals at a proportionally faster rate than we consume abundant metals.

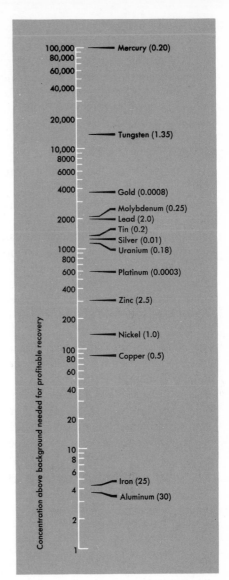

FIG. 6-2 The crustal abundances of geochemically scarce metals are so low that large concentrations above background average are needed before deposits can be profitably mined. The abundant metals require lower concentration factors to produce rich ores. The bracketed percentages are the minimum metal contents an ore must have before it can be mined under the most favorable circumstances with present-day technology.

In order to satisfy the growing demands, man has always sought *ore deposits*, the localized geological circumstances in which metals are sufficiently concentrated so that they can be won cheaply and rapidly. There is no reason to believe that the pattern will change in the future. A number of factors determines whether or not a local concentration can be considered an ore deposit. With all factors favorable, the present minimum concentrations are high (Fig. 6-2), although the minimums tend to change somewhat as mining becomes more efficient and as prices change. We do not yet know how low the necessary concentration factors can be pushed before the cost of recovering the metals is so high that we must do without, or before substitutes such as cements, ceramics, plastics, or even abundant metals will take their place. Technology too is a resource—of man's own ingenuity—and can be vitally important in evaluating potential resources of scarce metals. What is considered hopelessly impractical for exploitation by today's standards, or even by those projected for tomorrow, may well be tapped in the future because of suitable technological advances.

DISTRIBUTION OF THE SCARCE METALS

The scarce metals are widely distributed, but unlike the abundant metals they rarely form separate minerals. Instead, they reside in the structures of common rock-forming minerals, usually the silicates, an atom of a scarce metal substituting for an atom of an abundant element. For example, nickel atoms substitute for those of magnesium in olivine (Mg_2SiO_4), though commonly only to the extent of a few nickel atoms for every million atoms of magnesium. Substitution of foreign atoms causes strains in a mineral structure, and, accordingly, there are limits to the process; these are determined by temperature, pressure, and various chemical parameters related to the rock composition. For most common rocks and minerals, the limits are not exceeded and the scarce metals remain atomically locked in the host structures. To recover them, the host mineral itself must be broken down, and this is an expensive chemical process because silicate minerals are highly refractory and difficult to reduce. When the limits are exceeded, however, the substituting element forms a separate mineral, for example, pentlandite ($(Ni,Fe)_9S_8$), in the case of nickel; the way is then open for an inexpensive beneficiation process. The physical properties of pentlandite differ markedly from the associated silicate minerals, and simple crushing, followed by a concentration (based, for example, on density or surface property (flotation) differences), will produce an inexpensive, nickel-rich concentrate of pentlandite before the expensive chemical reduction process is started. The principle of having a scarce metal present in a separate mineral, which in turn has distinctly different physical properties from its associated minerals, has always been the most important factor in the utilization of the metals and the exploitation of their ores. It is significant, for example, that the scarce metal gallium, with a crustal abundance approximately twice that of lead, occurs almost exclusively as a substituting element and has never been of vital importance in technology. Had inexpensive sources of easily reduced gallium minerals been available, we would undoubtedly have myriad uses for them.

Only a tiny fraction of the scarce metals in the crust—less than 0.1 percent—occurs other than as atomic substitutes in common minerals. The substitution principle can sometimes be used to advantage, however. Elements such as silver and cadmium, which, because of low abundance levels, rarely form separate minerals are produced largely from ore minerals in which they are carried by atomic substitution. Silver, for example, commonly substitutes for copper in the ore minerals tetrahedrite ($Cu_{12}Sb_4S_{13}$) and chalcocite (Cu_2S), and it substitutes for lead in the mineral galena (PbS). Recovery of silver as a by-product is now so large that in 1972, of the ten largest silver producers in the United States, five were primarily lead and zinc producers, and three primarily copper producers, whereas only two mined their ores primarily for the silver content.

What Is an Ore Deposit?

An ore deposit is a mineral deposit from which one or more materials can be extracted profitably. Two essential features of mineral deposits have already been discussed: they are localized volumes within which certain chemical elements are concentrated far above the crustal average; they are places where special and unusual minerals occur. Mineral deposits only form when special processes occur; they are caused by rare circumstances within the crust. In addition to the scientific side of mineral deposits, there are two essential economic factors that must be satisfied if a mineral deposit is to be called an ore deposit. First, the deposit must be rich enough for profitable mining and large enough to offset the investment of a mining operation; second, the deposit must be geographically accessible so that transportation doesn't wipe out profits. The economic term profit is therefore an integral part of the definition of ore. There are no hard and fast rules to predict when a deposit contains ore and when it doesn't. For example, 10,000 metric tons of gold ore at a depth of 2,000 meters would probably be valueless unless it were adjacent to an existing mine and the cost of reaching it could be minimized. But the same material in the mountains of Antarctica would be valueless because of its remoteness. Each case must be evaluated on its merits, figuring in all costs of mining, smelting, and transportation. If metals can be produced competitively with other mines around the world, the deposit contains ore. If not, it is only a potential resource. Reserves and ore are therefore words that apply to the same material.

Oil pools, coal fields, and enriched cappings over banded iron formations are all examples of ores. Here we are concerned with deposits of geochemically scarce metals and their ores present special problems of their own. Scarce metal ore deposits tend either to have sharply defined boundaries like the walls of veins or to have gradational boundaries, as in some copper deposits, that grade into average rock over distances of tens, or at most a few hundreds, of meters. The deposits are rather like the raisins in a fruitcake. They are also small compared to deposits of abundant metals. For example, large copper deposits contain between 10^6 and 10^7 metric tons of copper, and the largest known does not exceed 5×10^7 metric tons. Many iron deposits, on the other hand, exceed 10^9 metric tons. The maximum size of scarce metal deposits and the number of large deposits of a given metal are both apparently related to the crustal abundance of the metals (Fig. 6-3(A) and (B)). The reasons for these relations have not been satisfactorily explained, but they are probably related to the degree of concen-

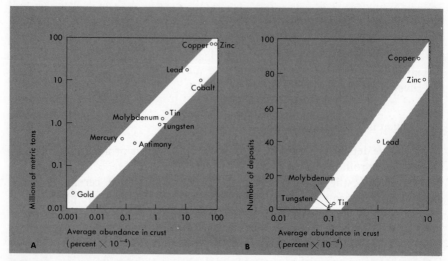

FIG. 6-3   (A) The largest known ore deposit for each scarce metal is approximately proportional to its crustal abundance. (B) The lower the concentration factor and the higher the crustal abundance, the more common it is for large ore deposits to form. The number of deposits containing a million metric tons or more of a scarce metal is proportional to the crustal abundance. Note that nonlinear exponential axes are used on both graphs.

tration needed for a scarce metal to be a minable ore. The scarcer the metal, the greater the needed concentration and the less likely it apparently is for all the special concentrating factors to be working at the same time and place.

Discovery of Ore Deposits

As new lands have been explored and prospected, they have followed patterns similar to those shown in Fig. 6-4. The number of ore deposits discovered, and therefore of working mines (curve A), climbs rapidly through an initial phase of prospecting. After a time, however, as smaller mines are worked out, the rate of discovery no longer exceeds the rate of depletion and closing. The number of working mines then declines and eventually must reach zero (although no large country has quite reached this point so far). During the active working life of the mines the production of metal increases (curve B), although lagging the mine curve because mines must be discovered before they can be worked. The metal production curve must eventually decline to zero too, when the last mine closes. In industrial countries in which mineral output is largely used for internal consumption, the inexorable rule of exponential growth soon makes it difficult for internal production to satisfy needs. Importation commences (curve C) and as time passes a growing amount of the metal consumed comes in from new lands abroad.

The curves in Fig. 6-4 are substantiated by recent history. Three prominent industrial countries are plotted in their relative positions. As Western man expanded into and explored the Americas, Africa, Asia, and Australia, he found abundant ore deposits. But in the more carefully prospected areas, such as the United States and Europe, the discovery rate has decreased drastically. It was recently pointed out by R. J. Forbes, for example, that the mines operating in the portion of Europe formerly embraced by the Roman Empire were all in mineralized areas known to and exploited by the Romans. The only exceptions were deposits of metals such as aluminum and chromium, metals that the Romans did not know how to use. Despite close settlement and keen prospecting by members of a society aware of the importance of metals, no totally new mineral districts were found during a period of almost 2,000 years.

As the inadequately prospected areas of the world diminish, we are forced to develop more sensitive means of searching beneath soil and overlying rock covers and to develop criteria for narrowing the areas of search to those few selected spots on the Earth where the probability of finding ores is greatest. How successful this will be is a great unknown. Many ore deposits tend to form at or within a few thousand meters of the Earth's surface. Scientific observation, therefore leads to predictions that the number of ore deposits waiting to be found will not increase with depth and will probably decrease. Hardly a comforting thought. We need more exact and more certain prospecting methods, particularly where buried ores are concerned. We also need much more work leading to the identification of unconventional potential resources

FIG. 6-4 Traditional stages in mine development, metal production, and imports in industrial countries. Curve A, which represents the number of working mines, rises rapidly as a new country is prospected, but it declines when the rate of mine exhaustion exceeds the discovery rate. Curve B, which represents metals produced, also rises and falls as mines are worked and eventually exhausted. Curve C, representing metals imported, rises exponentially and expresses the increasing inability of a country to meet its own needs. The approximate present positions for three industrial countries are indicated. With traditional development, each country moves along the time axis from left to right. For example, England was in about the position of the U.S.A. in the late nineteenth century, at which time the U.S.A. was about the same stage of development as the U.S.S.R. is today.

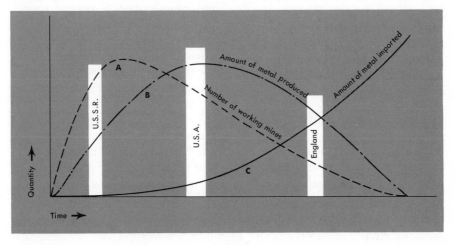

of scarce metals. Success or failure in these areas will directly determine our future use of scarce metals.

CLASSIFICATION OF SCARCE METALS

When scarce metals form ore deposits, the minerals in them have distinctive properties and compositions, and on this basis they can be grouped in three categories. The first, which includes copper, lead, and zinc, commonly forms *sulfide minerals*. The second, which includes gold and platinum, commonly occurs as *native metals*. The third, which includes tungsten, tantalum, tin, beryllium, and uranium, commonly forms *oxide* and *silicate minerals*. Some overlap occurs—tin, for example, forms both sulfide and oxide minerals—but the groupings take note of the major and most important minerals.

SCARCE METALS FORMING SULFIDE DEPOSITS

The number of scarce metals concentrated principally in sulfide deposits is large—copper, lead, zinc, nickel, molybdenum, silver, arsenic, antimony, bismuth, cadmium, cobalt, and mercury, together with numerous rarer ones that occur largely or solely as atomic substitutes for other scarce metals. Space demands that we discuss only the most important of these.

Copper

Copper, a metal used since antiquity, is now the workhorse of the electrical industry because of its excellent properties of conduction. Copper deposits are widespread, but mostly as veins or contact metamorphic deposits containing a million metric tons or less of copper, which is small for purposes of mining; they are often extemely rich, however, and many exceed 10 percent copper. Until the turn of the twentieth century, such deposits accounted for most of the copper produced in the world, and because of the high cost of working small deposits by underground mines, the lowest-grade ores that could profitably be mined were about 3 percent. Although copper is still mined on all continents, production from vein deposits has continually declined in importance as the discovery and exploitation of large, low-mining-cost deposits, such as the *porphyry copper deposits*, have proceeded.

Porphyry coppers are large, low-grade hydrothermal deposits containing at least 5 million metric tons of ore averaging 2 percent copper or less, in which the copper mineral, usually chalcopyrite ($CuFeS_2$) is so evenly distributed that large-volume, and consequently inexpensive, mining practices can be employed. The usual system employed is surficial mining from huge open pits (Fig. 6-5), and the very large volume has allowed the minable ore to diminish to today's low level of about 0.4 percent in some of the newest mines (Fig. 6-6). Porphyry

FIG. 6-5 Aerial photograph of the Morenci Open Pit from which the Phelps Dodge Corporation mines the porphyry copper deposit at Morenci, Arizona. The size of the pit can be gauged from the string of railroad cars carrying ore, visible in the left center of the photo, and from buildings visible on the left rim of the pit. (Courtesy of Phelps Dodge Corporation.)

copper deposits are typically associated with igneous intrusions that have a distinctive texture called porphyritic, which explains the name of the deposits, and which is characterized by large feldspar or quartz crystals set in a matrix of finer-grained minerals. The deposits are also characteristically contained in large volumes of rock that have been shattered, sheared, faulted, or somehow broken up on a fine scale, and through which mineralizing fluids have found easy passage. The shattering and mineralization may be in the intrusive, in the surrounding rocks, or both, and are apparently caused by the violent and often volcanic forces associated with in-

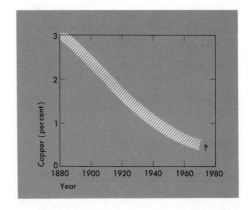

FIG. 6-6 A steady reduction in the minimum grade of copper ores that can be profitably worked has arisen from the discovery of large, low-mining-cost deposits and the development of ever larger and more efficient machines. Unfortunately, the same trend has not been observed for all metals because deposits amenable to mass mining methods have not been found. (After U.S. Bureau of Mines.)

trusion of the igneous rocks. Geologists are still unsure of all the events leading to formation of porphyry coppers, although the events have occurred repeatedly in some areas, such as the southwestern U.S. and the western Andes in Peru and Chile within the last 170 million years. Mineral deposits are commonly observed to be much more abundant, or even totally confined, within a limited geographic region. Such a deposit-rich region is commonly called a *metallogenic province* (Fig. 6-7). The distinctive metallogenic province within which the porphyry coppers are found in the Americas is so nearly parallel to the margins of the moving crustal plates that many geologists feel that a vital, but presently unexplained, connection must exist between them.

The first porphyry copper to be recognized is still worked at Bingham Canyon, Utah. It is now the largest copper producer in the U.S.A., but D. C. Jackling and R. C. Gemmell were treated with skepticism when in 1899 they first proposed the idea—revolutionary at the time—of bulk mining low-grade

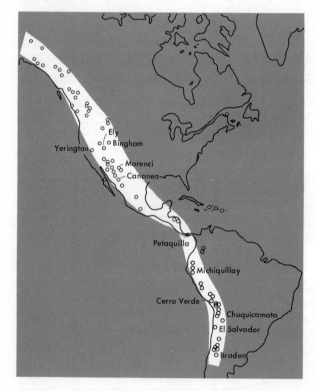

FIG. 6-7 Porphyry copper deposits in the Americas define a remarkable metallogenic province that lies parallel to the western continental boundary. Another belt of porphyry coppers is now being exposed by prospecting in the Pacific islands.

ores; but the idea proved successful. By 1907 a mill was built capable of handling the then huge tonnage of 6,000 metric tons per day. Capacity now exceeds 110,000 metric tons! More than 50 porphyry coppers have now been discovered; although principally clustered in the Americas, they have been discovered in the U.S.S.R., in Iran, in the Bor-Majdanpek area of Yugoslavia, and most recently in the Philippines, New Guinea, and adjacent Pacific islands.

A second type of large copper deposit, commonly called *stratiform* and believed by many to have formed as copper-rich chemical sediments, accounts for approximately 20 percent of the world's production of copper. Stratiform deposits—so called because they are confined to individual sedimentary horizons and conform in all details to the sedimentary layering—are known from several geological ages and have a wide geographic distribution. Each stratiform deposit has features somewhat different from others in the class, but since copper-rich sediments are not known to be forming today, direct observation cannot provide answers to the many puzzles about their origins.

The longest worked and most famous stratiform deposit in the world, and the one least altered by metamorphism—hence, the most informative—was laid down in Permian times in the area now occupied by central Europe. A sedimentary bed of organic-rich muds, rarely more than 60 centimeters thick, but now containing an unusually high concentration of copper, lead, and zinc sulfides, formed over an area of 50,000 square kilometers in a shallow sea of the time called the Zechstein (Fig. 6-8). The mud, now consolidated to a shale and called the *Kupferschiefer*, was not uniformly mineralized, for the copper ores are strongly concentrated in small areas in the western part of the basin—in East Germany and Poland. It is not certain that the sulfides precipitated from sea water contemporaneously with deposition of the muds, or whether they were introduced during the period immediately following burial, when many chemical and physical changes can occur in the enclosed organic and mineral matter.

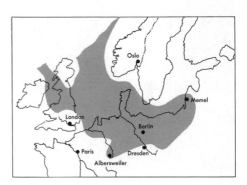

FIG. 6-8 Extent of the shallow Zechstein sea in which, during the Permian period, the thin sedimentary bed, now known as the Kupferschiefer was laid down. (After R. Brinckmann, 1960.)

There is a good deal of evidence to suggest that the original Kupferschiefer contained only fine-grained iron sulfide grains. Sometime after burial a brine carrying small amounts of copper, lead, and zinc chlorides apparently came into contact with, and replaced, the iron sulfides by the minerals we now find, much as an iron nail in a copper chloride solution is slowly transformed to a copper nail.

The Kupferschiefer has, after formation of the ores, only been slightly changed by low-grade metamorphism. There are also a number of large deposits in Precambrian sediments, and most of these have been more intensively metamorphosed, making evidence of their origins even more ambiguous. In the United States the most important deposit of the type—in the White Pine area of Michigan—has many characteristics in common with the Kupferschiefer, although the evidence at White Pine clearly proves a secondary or postdepositional origin. Of considerable importance too are the deposits in the Dzhezhazgan-Karsakpay area of Kazakhstan, but the most remarkable of all occur in a belt of exceedingly rich deposits known as the *Zambian Copper Belt*. These extraordinary deposits occur in sediments that were deposited along ancient shorelines about 700 million years ago. The deposits form an elongate belt (Fig. 6-9) that straddles the Zambian-Zaire border and is one of the most richly mineralized regions in the world. As with other stratiform copper deposits, those of the Zambian Copper Belt are surrounded by a controversy concerning their origins; some experts favor a sedimentary origin, others a secondary origin.

More than 50 percent of the world's copper production comes from porphyry coppers, the largest share coming from the U.S.A., and an estimated

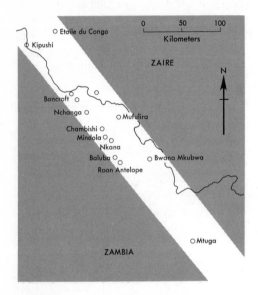

FIG. 6-9 The Zambian Copper Belt, a remarkable series of rich copper deposits in Zambia and Zaire, contained within sediments laid down along an ancient Precambrian shoreline. Scientists are split in their opinions as to whether the copper minerals were deposited with the sediments or were introduced later. (After F. Mendelsohn, 1961.)

20 percent from stratiform deposits, principally those in Zambia. The remainder comes from a great variety of small deposits that are widely distributed and that afford most countries at least a small production. In 1972, 60 countries produced copper and sold it on the world market. Because of the overwhelming importance of the porphyry and stratiform deposits, however, 8 countries account for fully 80 percent of the world's production (Table 6-1).

With a current annual production in excess of 6 million tons of copper and with rapidly rising consumption curves, we must ask what the future might be for copper. Reserves are large, as shown in Table 6-1, but with a consumption doubling period of about 15 years, the world will need about 400 million metric tons of new copper by the year 2000. What, then, are the potential resources?

Table 6-1 Leading Copper-Producing Countries and Their Reserves, 1971

Country	Production (metric tons)	Reserves (metric tons)
U.S.A.	1,381,000	77,000,000
CHILE	716,000	53,000,000
CANADA	653,000	9,000,000
ZAMBIA	651,000	27,000,000
U.S.S.R.	617,000	34,000,000
ZAIRE	407,000	18,000,000
PERU	213,000	23,000,000
PHILIPPINES	208,000	9,000,000
OTHER COUNTRIES	1,198,000	36,000,000
TOTAL	6,044,000	286,000,000

(After U.S. Bureau of Mines.)

They are exceedingly difficult to evaluate because we must guess what deposits might be found in the future, and as we have seen, the past record is not encouraging unless new country can be prospected. New deposits will, of course, be found, and some existing ones will prove larger than presently thought. Even so, many specialists propose that potential resources in conventional deposits might be no greater than 500 million metric tons of copper. By rough rule of thumb, this is equivalent to finding about as many new deposits as we have already found. Furthermore, the richest deposits might lie in the vast spaces of the U.S.S.R. east of the Urals and in South America.

Potential resources of copper from unconventional sources are even more difficult to evaluate. The copper contents of manganese nodules in the deep oceans are sufficiently high—in places as much as 2 percent—so that the nodules may reach the status of ore within the present decade. Estimates of total recoverable tonnages from nodules are little more than wild guesses, but if the estimated 2×10^{12} metric tons of nodules on the floor of the Pacific yield 0.5 percent copper, the total would be 10^{10} metric tons—larger than all the conventional reserves and potential resources. Other potential resources are less obvious. Below contents of about 0.1 percent, copper tends to occur by atomic substitution in silicate minerals. Continual lowering of the mining grade, therefore, will bring us to a barrier that will require radical new technologies and vast supplies of energy to surmount. The curve in Fig. 6-6 seems to have a limit, and its projection is not something in which to place our hopes for the future. The prediction for copper is therefore uncertain. Rising prices may make it possible to surmount the technologic and energy barriers perceived ahead, but the prognostication of uncertainty for copper is not unique. Most scarce metals suffer from the same problem.

Lead and Zinc

Lead and zinc are discussed together because their ore minerals occur together, and in each case a single mineral species, galena (PbS) and sphalerite (ZnS), respectively, accounts for most of the world's production. Lead is used principally for storage batteries and for lead tetraethyl, an antiknock additive to gasoline, but it also finds wide usage in the construction industry. Zinc, used principally as a component of brass until the eighteenth century, has many uses today, but more than 50 percent of production is consumed in the preparation of alloys for die-cast products and in anticorrosion treatment of iron and steel.

Like copper ores, lead and zinc deposits occur in two very different ways: as hydrothermal and contact metamorphic deposits and as stratiform deposits. In both cases the ores tend to be rich but confined in size so that costly underground mining procedures are usually necessary for recovery. Large, low-grade types of deposits analogous to the porphyry coppers have not yet been discovered.

Hydrothermal deposits are widespread and are usually closely associated with igneous intrusions, but one particularly unusual variety is an exception to the rule; it is known as the *Mississippi Valley type*, after the remarkable metallogenic province stretching from Oklahoma and Missouri to southern Wisconsin and coinciding with much of the drainage basin of the Mississippi River system. Deposits of similar affinity have been identified in several parts of Europe, northern Africa, northern Australia, the U.S.S.R., and, most recently, Pine Point in Canada's Northwest Territory. This last is quite possibly the largest deposit ever discovered, and it may belong in the same metallogenic province as the deposits in the United States.

Mississippi Valley deposits occur principally as replacement bodies in limestones of many ages. Solutions carrying the scarce metals apparently dissolved the limestone and slowly deposited the galena and sphalerite that often form large and beautiful crystals (Fig. 6-10). The deposits are usually far from any obvious igneous activity, and a controversy of long standing surrounds the source of solutions. In recent years it has been established, from the analysis of tiny fluid samples trapped as inclusions in imperfections of the growing crystal (Fig. 6-11), that the solutions were brines with close affinities to those from some oil fields. They are also similar to brines encountered in certain geothermal reservoirs, most notably one that occurs in the the region of the Salton Sea in the Imperial Valley of Southern California. The waters have their origin on the surface, but as they penetrate deep underground they are heated and commence reacting with minerals in the enclosing reservoir rocks. The reservoir rocks recrystallize and in the process the trace amounts of scarce elements locked up by atomic substitution are released to the brine. The brines are therefore metamorphic hydrothermal solutions even though the heat that caused them may have been derived from an igneous rock. As the brines move out through openings and channelways depositing minerals and forming ore deposits, they are nearly indistinguishable from igneous hydrothermal solutions.

FIG. 6-10   Large, well-formed crystals of galena (PbS) from the Mississippi Valley type deposits near Picher, Oklahoma suggest slow growth and absence of the frequent shattering characteristically associated with many hydrothermal deposits. The cubes of galena in this photograph are 5 centimeters across.

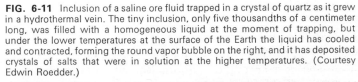

FIG. 6-11 Inclusion of a saline ore fluid trapped in a crystal of quartz as it grew in a hydrothermal vein. The tiny inclusion, only five thousandths of a centimeter long, was filled with a homogeneous liquid at the moment of trapping, but under the lower temperatures at the surface of the Earth the liquid has cooled and contracted, forming the round vapor bubble on the right, and it has deposited crystals of salts that were in solution at the higher temperatures. (Courtesy Edwin Roedder.)

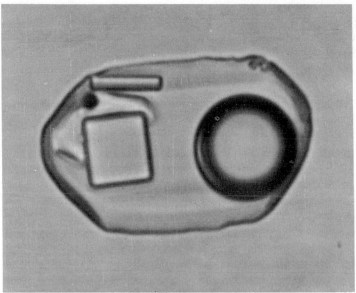

Another class of lead and zinc ores formed by hydrothermal solutions is found enclosed within lavas erupted on the sea floor. They are commonly massive, always comprised of sulfides, and commonly rich in copper minerals as well as minerals of lead and zinc. The origin of massive sulfide deposits poses yet another mystery. Because they are enclosed within igneous rocks, there can be no doubt that an igneous origin for the depositing solutions is possible. Yet the association with submarine volcanism, and a number of suggestive clues within the ores themselves, leads many to suspect that the depositing solutions may be metamorphic fluids formed by sea water circulating through and reacting with vast piles of hot lavas, in the process picking up trace metals from deep within the piles and releasing them as they return and mix with cool sea water. Perhaps similar deposits are somewhere forming today on the sea floor and oceanographers might therefore be the ones to solve the puzzle. One area of lead, zinc, and copper sulfides has already been discovered. In the Red Sea, which marks a line of active volcanism along a join where two crustal plates are moving apart, three places have been found where hot, metal-rich brines debauch into sea floor basins. The brines deposit sulfide minerals and are perhaps forming what might, in future geological ages, be a massive sulfide deposit.

An increasingly large production of lead and zinc ores has come from stratiform deposits, and as with the copper ores, origins are controversial. The clearest example of a stratiform deposit is again the Kupferschiefer, where the deeper parts of the Zechstein basin apparently favored the accumulation of lead and zinc sulfides rather than of copper. Although the ores are low in grade, the Kupferschiefer has been worked for both lead and zinc in many places. Like copper deposits, most of the stratiform lead-zinc ores are Precambrian and are now so changed by metamorphism that evidence of their origin has been obliterated. One very large deposit, the Sullivan mine, occurs in Canada, but the three largest deposits of the class are all in Australia. The first, called the H.Y.C. deposit, is essentially unmetamorphosed and occurs in the far northern part of the country near the McArthur River. The H.Y.C. has many features of a primary sediment, but close examination again shows that the ores could just as well have arisen by reactions between circulating brines and a black shale. The second deposit is at Broken Hill in the arid western part of New South Wales and the third at Mt. Isa, several hundred kilometers to the north in Queensland. Broken Hill and Mt. Isa are not only two of the larges ore deposits known, they are also among the richest. At Broken Hill, ores containing more than 20 percent each of lead and zinc are known.

The world's production of both lead and zinc is dominated by the same five countries (Table 6-2). The reserves reported for both metals are probably not accurate measures of metal available. Some countries, such as the U.S.S.R., do not make all information available, and in many mines the cost of making accurate measurements of ore that will only be mined many years hence is not condoned. Potential resources of quality and quantity equal to the reserves will probably be developed in the course of normal mining development in the existing deposits. Beyond this point, however, the situations for both lead and zinc are similar to that for copper—with a marked exception. Potential resources of low-grade and unconventional materials have not been identified.

Table 6-2 Leading Lead- and Zinc-Producing Countries and Their Reserves, 1971

	LEAD		ZINC	
Country	Production (metric tons)	Reserves (metric tons)	Production (metric tons)	Reserves (metric tons)
U.S.A.	525,000	32,000,000	452,000	30,000,000
U.S.S.R.	446,000	4,500,000	645,000	12,600,000
AUSTRALIA	395,000	9,000,000	445,000	8,000,000
CANADA	384,000	10,800,000	1,257,000	22,500,000
PERU	175,000	4,500,000	331,000	7,000,000
WORLD TOTAL	3,376,000	85,000,000	5,470,000	111,400,000

(After U.S. Bureau of Mines.)

Nickel

Nickel, used almost entirely as an alloying metal in the production of special products such as stainless steel and high-temperature and electrical alloys, is a product of twentieth-century technology. Smelting and working the metal is so difficult that old German miners, who confused the similar-looking copper and nickel sulfide ores, called it kupfernickel after "Old Nick," who supposedly bewitched the copper (kupfer) ore and thus made it impossible to handle. The frustrations of these miners live on in our use of the word nickel.

There are only two important ways in which nickel is recovered. The first, accounting for the large Canadian, Russian, and Australian productions (Table 6-3), is from the mineral pentlandite, which commonly occurs in deposits formed by magmatic segregation.

Table 6-3 Leading Nickel-Producing Countries and Their Reserves, 1971

Country	Production (metric tons)	Reserves (metric tons)
CANADA	267,000	9,000,000
U.S.S.R.	118,000	9,000,000
NEW CALEDONIA	102,000	15,000,000
CUBA	36,000	16,000,000
AUSTRALIA	31,000	1,000,000
WORLD TOTAL	640,000	68,000,000

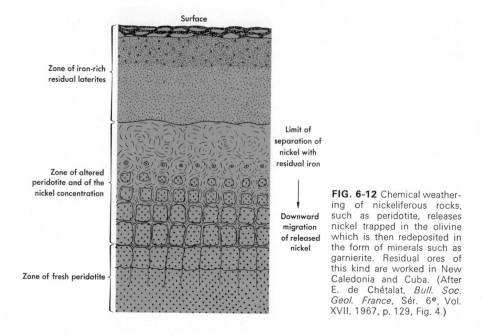

FIG. 6-12 Chemical weathering of nickeliferous rocks, such as peridotite, releases nickel trapped in the olivine which is then redeposited in the form of minerals such as garnierite. Residual ores of this kind are worked in New Caledonia and Cuba. (After E. de Chétalat, *Bull. Soc. Geol. France*, Sér. 6ᵉ, Vol. XVII, 1967, p. 129, Fig. 4.)

The second is in the residual weathering zones formed over certain igneous rocks in tropical regions. We previously used the example of nickel substitution in an olivine. Weathering releases the trapped nickel to surface waters, and under some conditions it is reprecipitated as relatively insoluble nickel silicates (Fig. 6-12), such as garnierite ($H_4Ni_3Si_2O_9$), named for the Frenchman Garnier who first discovered the rich nickel ores of this type in New Caledonia. Beneficiation, which is so costly to man, has thus been effected by nature in a slow, long-continued weathering process.

Although quite pure deposits of garnierite may be formed in this fashion, it is more common for tropical weathering to produce an iron-rich laterite in which the concentration only reaches a level of about 1 percent nickel. Low-grade deposits of this kind are widespread in the tropics, and nickeliferous laterites measured in the hundreds of millions of metric tons have been discovered in Cuba, the Philippines, Greece, Borneo, and other parts of the world. Though only worked in Oregon in the United States and in New Caledonia and Cuba at the present time, the laterites constitute man's largest known potential resource of nickel, and together with the large segregation deposits they appear to be adequate to meet our demand for nickel far into the future.

Molybdenum

Molybdenum, like nickel, is a metal used mainly as an alloying element in steels, to which it imparts toughness and resilience. Molybdenum first gained widespread importance when it was used in steels for armor plate and armor-piercing shells during World War I. It now has a wide variety of uses, particularly in alloys in which resistance to wear and retention of strength at high temperatures are required.

One mineral, molybdenite (MoS_2), accounts for most of the world's production of molybdenum, but because it has a highly erratic distribution it is difficult to estimate accurately the world's resources. A significant fraction—approximately 25 percent of the world's current molybdenum production—is derived as a by-product of porphyry copper mining. The content of porphyry copper ores is low—in the range 0.01 to 0.04 percent molybdenum—but the tonnages of ore processed are so large, and the costs of concentrating molybdenite by flotation so low, that a profitable recovery is effected at many of the larger porphyry copper mining operations.

The site of the major production and future resources of molybdenum is a series of deposits that have many geological similarities to porphyry coppers and are often called stockwork molybdenum deposits (because the molybdenite is confined to a myriad of tiny fractures that form a stockwork throughout the porphyry). The deposits are found in a metallogenic province stretching from the Mexican border to northern British Columbia (Fig. 6-13).

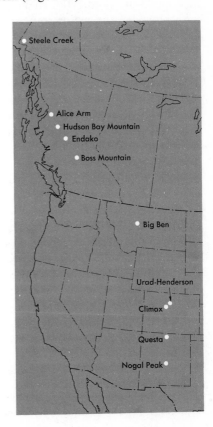

FIG. 6-13 Distribution of stockwork molybdenum deposits in a metallogenic province that falls near the eastern margin of the Cordillera of North America. (After K. F. Clark, 1968, *Economic Geology*, v. 63, p. 560.)

One of the deposits, at Climax, Colorado, has supplied half of the world's production for the last 50 years. Such heavy dependence on a single ore deposit is unwise, of course, and the situation is now eased by the commencement of large-scale mining operations at other deposits in Canada and the United States.

The estimated current world production rates of molybdenum are approximately 78,000 metric tons per annum. With reserves of 2,860,000 metric tons of molybdenum in the U.S.A. alone, we are unlikely to face a resource problem with molybdenum supplies for many years to come.

Silver

The widespread attention paid to silver shortages by newspapers is derived from its use in currency and monetary speculation. A real shortage is arising from industrial consumption, however, with the photographic and electrical industries consuming an ever larger share of a supply that cannot be easily expanded to satisfy demands.

Shortly after Columbus blazed a pathway to the Americas, the great silver bonanzas of Central and South America were discovered. To the present day, deposits in the Cordilleran chain, stretching from Alaska to Tierra del Fuego, have remained the major source of the world's silver supply. Approximately 55 percent of the silver produced annually comes from countries in the Americas, with most of the remainder coming from Australia and the Soviet Union, and only marginal production from Asia and Africa.

Silver minerals commonly occur in hydrothermal vein deposits, and characteristically, either they are associated with lead, zinc, and copper minerals, or else the silver itself is carried in the lead and copper minerals by atomic substitution. Only a minor percentage of silver-producing deposits are rich enough to be worked for silver alone, however, and therein lies the silver problem. Most silver is produced as a by-product from copper, lead, and zinc mining, and its production rate is controlled strictly by the production rates of the associated metals. Although a few new silver deposits have been found in recent years, so much silver is produced as a by-product that silver production has not been able to expand to meet the growing need. With the 1965 decision by the United States Government to withdraw silver from currency, the price of silver has climbed steadily, and it may be anticipated that further increases are ahead unless large new supplies are found. Such discoveries do not seem likely. Though old mines may be reactivated and formerly uneconomic ores become profitable in the face of rising silver prices, there seems little hope that silver production will grow to meet all demands; rather, it is likely that future uses will have to be curtailed in order to meet the limited supply.

Other Sulfide-Forming Scarce Metals

The remaining scarce metals that occur principally as sulfides do not warrant separate discussion. Almost without exception, small deposits are widespread and resources seem adequate for foreseeable future needs. The one apparent exception is mercury.

Most of the world's mercury production comes from one mineral, the vermillion-colored cinnabar (HgS), which is found erratically distributed in narrow hydrothermal veins in a number of volcanic areas. The known deposits are all shallow, and most are so small that they have been exhausted soon after discovery. The world's current production comes largely from Spain—where the Almadén mine has been producing for more than 2,000 years, from Idria in Yugoslavia, and from Italy, all regions of extensive Cenozoic volcanism. The once great deposits of California and Nevada are apparently almost exhausted. Known reserves of mercury are small and its by-production from other mining activities are limited. Many experts now suspect that it will be the first scarce metal to exhaust available supplies, and that this day will be reached in the twentieth century.

SCARCE METALS OCCURRING IN THE NATIVE STATE

Platinoid Metals

The scarce metals occurring in the native state are a less diverse group than the sulfides. Platinum, palladium, rhodium, iridium, ruthenium, and osmium, collectively called the *platinoid* elements, always occur together and are concentrated in igneous rocks derived from mantle magmas. The abundance levels of the platinoids are all low—platinum and palladium, the most abundant of the group, are only present in the crust to an extent of 0.0000005 percent; the others are even less abundant. However, the platinoid abundance in mantle rocks is noticeably higher though not sufficiently so for the rocks to be considered potential resources of platinoids without an additional concentrating factor coming into play. This occurs when magmatic segregation of nickel, copper, and iron sulfides occurs. As the compounds form and sink to the floor of the magma chamber, they act as atomic collectors and sweep the platinoids out of magmatic solution. In mining sulfide segregation deposits for copper and nickel, therefore, platinoids are recovered as valuable by-products. Within South Africa's Bushveld Igneous Complex there is a segregation layer called the Merensky Reef, which, although only a few tens of centimeters thick, is so rich that it can be mined for its platinoid content rather than for the nickel and copper it contains.

There is one other important occurrence of platinoids. The metals are essentially unaffected by chemical weathering and are malleable, so grains do not disintegrate as they are transported and deposited in sediments; being very dense, they concentrate in placers (Fig. 6-14). Where the sediment source area contains suitable igneous rocks, as in the Ural Mountains in the Soviet Union, important placer concentrations of platinoids develop.

Reserves of platinoid elements are quite large compared to the small production (Table 6-4) but they are very unevenly distributed, most being in

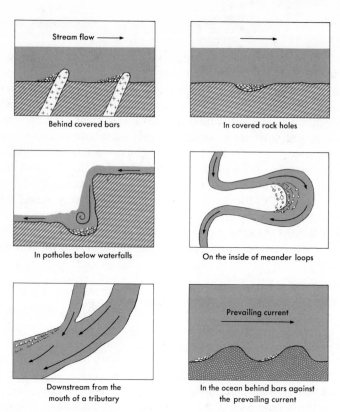

FIG. 6-14 Typical sites for placer accumulations which occur where obstructing or deflecting barriers allow faster-moving waters to carry away the suspended load of light and fine-grained material while trapping the more dense and coarse particles which are moving along the bottom by rolling or only partial suspension. Placers may form wherever moving water occurs, though they are most commonly associated with streams.

Table 6-4 Leading Platinoid- and Gold-Producing Countries, 1971*

| Country | PLATINOIDS | | GOLD |
	Production (metric tons)	Reserves (metric tons)	Production (metric tons)
U.S.S.R.	71.5	6,200	208
REPUBLIC OF SOUTH AFRICA	38.9	6,200	976
CANADA	14.6	500	70
U.S.A.	0.6	90	46
AUSTRALIA	—	—	21
WORLD TOTAL	126.8	?	1,446

* The platinoid elements are grouped because of similar occurrences and uses. Within the group, platinum and palladium each account for about 40 percent of the production and reserves, rhodium 9 percent, iridium 6 percent, ruthenium 4 percent, and osmium 1 percent.

(After U.S. Bureau of Mines.)

the Bushveld Complex and in the U.S.S.R. Potential resources are uncertain, but they are probably large. Most layered intrusives, such as the Stillwater Complex in Montana and the Duluth Gabbro in Minnesota, have large but low-grade potentials. The main trouble with platinoids is that production of very low-grade resources may have to be linked to production of the copper and nickel. Thus, the future may see a platinoid situation analogous to the present one for silver.

Gold

Gold, unlike the platinoids, is not commonly associated with mantle rocks, but it is characteristically associated with igneous rocks formed from crust derived magmas. Like sulfide minerals, gold is apparently transported by hydrothermal solutions, though how this is done remains a point of conjecture, for native gold is one of the least soluble substances known; it is commonly found in hydrothermal vein deposits, either associated with sulfide minerals or alone. Very rarely does an ore contain much gold; a rich ore, for example, contains only 0.007 percent gold and yields about 60 grams of gold for every metric ton of rock mined. Much leaner ores than this can be profitably worked, however.

Gold is so resistant to corrosion that it is almost indestructible. Most of the gold ever mined is still available, having been used and reused many times in its passage through history. Some of the gold from Cleopatra's bracelets, for example, may reside in your tooth fillings or wedding band. Gold is also an unusual element in that it forms a separate mineral even at very low concentration levels and is apparently not concealed by atomic substitution to the same extent that many other, even more abundant, elements are. Because gold is widespread in tiny amounts and is very dense and indestructible, it is ideally suited for concentration in placer deposits, from which has come much of the world's production.

Among the most remarkable of all the mineral deposits in the world are the gold deposits of the Witwatersrand district of South Africa. Formed in Precambrian times, the ores are contained in a series of conglomerates—sedimentary rocks consisting of rounded pebbles cemented by a finer-grained matrix. This is exactly the rock type that commonly carries placer concentrations in most parts of the world and in all geologic ages; there the story would end but for the astonishing extent of the Rand ores. Most placer deposits are small, perhaps a few hundred or thousands of meters in extent, and are clearly confined to present or former narrow stream channels. The Rand ores too are apparently localized by what were old channels, but the deposits have been mined along an outcrop length of conglomerates in excess of 400 kilometers, and they have been followed to the limits of practical mining—more than three and a half kilometers deep. Although most who have studied these are hesitant to propose alternatives to placer deposition for the Rand ores, no one has successfully solved the problem of their extent, nor of their apparent uniqueness.

First discovered in 1885, the Rand district soon became, and has remained, the world's leading gold-producer. In 1973 it accounted for more than 67 percent of the world's total production and a significant fraction of the uranium, which is produced as a by-product. Although gold production is widespread in small amounts around the world—71 countries have recorded some production—large deposits are rare, and only 5 countries account for 90 percent of production. The largest gold mine in the United States is the Homestake Mine in South Dakota. The deposit is Precambrian in age, has been metamorphosed, and is of uncertain origin. Scientists who have examined it suggest that hot spring waters deposited gold and associated scarce metals into the sedimentary host rocks as they were being deposited, and that further concentration into veins and pods occurred during metamorphism. Three deposits of an unusual type have recently been discovered in Nevada, the Carlin, Cortez, and Getchell mines, giving some hope that more deposits remain to be found. These deposits are hydrothermal and the gold is so fine-grained that it can rarely be seen even under a microscope. Finding new deposits, therefore, will call for very skilled prospecting.

In addition to its widespread monetary use, gold has other uses that continually drive the consumption rate upward. The largest use is in jewelry, but electronic components for computers are rising rapidly and dental needs present a steady demand. For many years the price of gold was controlled by governmental edicts, but with the establishment of a free world market in 1968, and restoration of rights of private ownership within the United States, the price of gold moved rapidly above $5 a gram. The question of reserves and potential resources must be viewed in light of a rising price. In 1968 the U.S. Bureau of Mines estimated that world reserves of gold ore that could be profitably worked at a price of $4.66 a gram ($145 an ounce) contained approximately 37,000 metric tons of gold. The price of gold soon rose above $4.66 a gram, but world reserves did not increase. Nor will further price rises probably increase the reserve figure greatly, because large, low-grade potential resources of gold have not been identified. Thus, gold may present something of an economic anomaly. The price may rise, but the amount produced may not increase at the same rate.

SCARCE METALS FORMING OXIDE DEPOSITS

With the exception of tin, the scarce metals with affinities for oxygen are all newcomers, and the diverse uses to which they are put reflect the amazingly complex technology we now support. Chromium, tungsten, tantalum, vanadium, and niobium (formerly known as columbium) are principally used as alloying agents in special steels, but they also have other highly specialized uses. For example, 33 percent of the tungsten produced is now used in the manufacture of tungsten carbide (WC), an extremely hard substance used for cutting edges in metal work, in rock drills, and in armor-piercing shells; tantalum has been proven to have the special electronic properties needed for low-power valves, and it offers revolutionary possibilities for computers of the future. Uranium, of course, has become the wonder metal of our age now that atomic energy can be released by controlled fission. Because its uses are those of a fuel rather than a metal, uranium was separately discussed in Chapter 4.

Chromium

Chromium, the oxide-forming scarce metal with the largest annual production, is an essential alloying metal used in the steel industry, and one for which satisfactory substitutes have never been found. Important amounts are also consumed in the chemical industry and in the use of chromite $(Mg,Fe)_2CrO_4$, for refractory bricks.

Until the end of the last century, chromium was used only in the chemical industry, principally for pigments. After 1900, however, it gained in importance as an alloying element; high-speed tool steels and stainless steels with a chromium base find ever wider application. Chromium is now an essential metal of modern technology.

Chromium is only known to form a single ore mineral, chromite. Supplies of chromite are limited, both in abundance and in geographic diversity. Alternative sources of chromium have not been identified.

Chromite deposits form only by magmatic segregation; chromite lends itself to concentration in this manner because it is very dense and crystallizes early in the magmatic cooling cycle. Unfortunately, the kinds of magma in which chromite forms are characteristic of the mantle and are rare in the crust. Deposits are found, therefore, where fragments of the mantle reach the surface or where mantle-generated magmas invade the crust. This happens in two ways. First, fragments from the upper mantle are sometimes torn off and thrust upward along the sutures where the moving plates of crust come into colliding contact. These mantle fragments, commonly called *peridotites*, are igneous rocks and they sometimes contain layers and pods of chromite formed by magmatic segregation. Because peridotite masses are found in mountain belts they are called *alpine peridotites*. The second way chromite reaches the crust is as a constituent of magmas. We now find the magmas cooled into layered intrusives, and the chromite is concentrated in rich magmatic segregates.

Much of the world's past production of chromite has come from alpine peridotites in the Ural Mountains of the U.S.S.R., the mountains of Turkey, the Philippines, Greece, Iran, Yugoslavia, India, and the United States. Unfortunately, the deposits are all small, difficult to locate, and in many cases difficult to mine. Increasingly, therefore, attention has been paid to layered intrusives, and here the reserves are much larger. But most of them are located in Africa, especially South Africa, where one of the most unusual masses of igneous rocks in the world occurs. Known as the Bushveld Complex, and comprised of many different layers of igneous rock, several of which contain chromite, the Complex covers hundreds of square kilometers and contains reserves of 10^{10} metric tons. Another large layered intrusive, called the Great Dike, occurs in Rhodesia, and here too vast reserves of chromite are known. It is clear that southern Africa will become even more important in the production of chromite than it now is. Eventually this part of the world will dominate the trade, supplanting the traditional leader, the U.S.S.R.

Fortunately, layered intrusions are found on all continents, and most contain low-grade potential resources of chromite. Many countries could, if they wished, be self-sufficient. There does not seem a danger of a chromium shortage, therefore, but rather a period ahead when the cost of chromium could rise as lower grade, but still abundant, resources are developed.

Tin and Tungsten

Tin is the only scarce metal of the group besides chromium with a large production, and it has been used as an alloying agent in bronze for thousands of years. The largest uses are now in the tinplate industry, where tin serves as an anticorrosion coating on iron and steel and in the production of soft solders.

Tin and tungsten minerals often occur together. They are found concentrated adjacent to igneous rocks in the continental crust, either in contact metamorphic deposits or as minerals in hydrothermal veins. Their most common minerals—cassiterite (SnO_2), wolframite ($FeWO_4$), and scheelite ($CaWO_4$)—are dense, erosion-resistant minerals that are readily concentrated in placer deposits. Much of the world's supply, particularly of tin, has come from such secondary concentrations.

Tin and tungsten are both elements for which the problems of future resources loom large; their abundances are low and their deposits scarce. North America, for example, is almost devoid of known tin deposits, though it is more fortunate in its tungsten resources. Most of the world's tin production and the known reserves and potential resources are concentrated in two narrow metallogenic provinces, one that runs first along the Malayan Peninsula, then southeast to Java and to Indonesia, and another that runs along the eastern side of the high Andes, in Bolivia and Peru. The major tungsten production and reserves are also restricted geographically, with approximately half of the world's production coming from an east-Asian belt running from Korea, through southern China, to Malaya.

Other Oxide-Formers

There are a number of other oxide-forming scarce metals, but the most important are niobium, tantalum, and beryllium, all of which are found concentrated in a special, rare type of igneous rock called a *pegmatite*. Pegmatites are small—rarely more than a few tens of meters in their longest dimension—and are associated with quartz-rich igneous rocks crystallized at great depth. It is believed that they are formed by the water-saturated residue left after crystallization of the bulk of the magma, and that they only form under deep-seated, high-pressure conditions where water and other volatiles cannot readily escape to form hydrothermal deposits. Elements such as niobium and beryllium, which do not readily substitute in the earlier-formed silicate minerals, become concentrated in pegmatites. Because the pegmatites form at considerable depth in the crust, they are only exposed in deeply eroded terrains.

For many years pegmatites served as the only source for niobium, tantalum, and beryllium. Within recent years, however, new and unexpected sources have been found for each metal. Beryllium silicate has been found as an exceedingly fine-grained and previously overlooked mineral in certain hydrothermally altered lavas in Utah, Nevada, and New Mexico. Although the beryllium mineral is too fine to allow beneficiation, it can be readily leached and recovered by acid solutions. As a consequence the Spor Mountains deposit in Utah has become the largest potential resource of beryllium for the future.

Niobium and tantalum minerals also occur in an unusual igneous rock called *carbonatite*, the origin of which also remains somewhat of a mystery. Carbonatites are intrusive rocks composed largely of calcium carbonate. Most rocks rich in calcium carbonate are sediments formed at the Earth's surface that contain a distinctive array of minor elements. Carbonatites, however, contain an unusual and very different array of minor elements, including niobium and tantalum; this strongly suggests that they are not generated by any melting process involving deeply buried carbonate sediments. Rather, they seem to be derived by some presently unknown means of differentiation from other igneous rocks.

SCARCE METALS IN THE FUTURE

Two points about the scarce metals raise uncertainty for the future. The first concerns the occurrence and distribution of ore deposits—they are limited in size, geographically restricted, and difficult to find. Evaluating potential resources, we find that large, low-grade and unconventional deposits have not been identified for most metals. We cannot even be sure whether or not deposits of any kind are likely to occur in the composition gap between presently minable ores and average crustal rock composition. In the case of copper, unconventional resources will be available when the presently exploited deposits are used up. But for metals such as tin, tungsten, and mercury, no alternatives are yet in sight.

The second point concerns use patterns. We use scarce metals at a much faster rate than is prudent, considering their scarcity. Balanced use of minerals proportional to abundance indicates increased usages of iron, aluminum, and the other abundant metals, and it indicates decreased use rates for scarce metals. For example, if we take the present annual world consumption of iron, about 475 million metric tons, as a base and use other metals in proportion to their geochemical abundances, the consumption rate of copper would be about 475,000 metric tons, of lead, 82,000 metric tons, and of mercury, 165 metric tons; in fact, the consumption rates are respectively 13, 41, and 91 times greater.

seven

fertilizer
and
chemical minerals

. . . the use of chemical fertilizer in 1960 was about six times the level of the 1930's. The additional production flowing from the recent rate of consumption of some 6 to 7 millions tons of plant nutrients per year . . . is roughly equivalent to the yield from some 75 million acres of unfertilized farmland. (H. H. Landsberg, Natural Resources for U.S. Growth, 1964.)

The term *nonmetallic minerals* is somewhat ambiguous, being neither a strict scientific nor an exact economic term. It is widely used, however, and embraces a group of minerals used for purposes other than the metals or other elements they contain. Unlike the metals, the nonmetallic minerals cannot be readily classified in terms of crustal abundances; they can, however, be simply classified on the basis of use. First, there are the minerals of primary use for fertilizers and for raw chemicals, accounting for 33 percent of the value of the nonmetallic production. Second, there are the materials for the building and construction industries discussed in Chapter 8.

MINERALS FOR FERTILIZERS

The fertilizer minerals are one of the most vital resource needs because they are essential for increased food production to meet the demands of the expanding population. Plant growth requires many elements. Oxygen and hydrogen (both derived from water) together with carbon (derived from atmospheric carbon dioxide) make up 98 percent of the bulk of the living plant. But nitrogen, phosphorus, potassium, calcium, and sulfur are also essential, and for the land plants—the source of our food supplies—they are provided by the soil. The rate of supply in part determines the rate of plant growth, and, to have any effect at all, the elements must be supplied in a form that the plant can assimilate. For example, most soils contain 1 percent or more of potassium, but to the plant, which needs soluble material, it is unavailable because it is locked in insoluble silicate minerals such as feldspars. Therefore, to enhance growth rates by addition of a potassium fertilizer, we usually add the potassium as the chloride (KCl), the most abundant soluble potassium mineral. The efficacy of the nitrogenous, phosphatic, and other fertilizers similarly depends on their solubility, and it is in the form of soluble compounds, or minerals that can be rendered soluble by minimal treatment, that fertilizer resources are sought.

The most essential fertilizers—the big three—are phosphorus, potassium, and nitrogen. They are applied to soils in the approximate weight ratios 1 to 1.5 to 3. Demand has been climbing very rapidly (Fig. 7-1) so that the total world fertilizer consumption is doubling on a 10-year cycle. If the population continues to rise and is to be fed, there seems little chance of fertilizer demands declining in the foreseeable future. Fortunately, the reserves are very large, but with the exception of nitrogen, reserves of presently used materials suffer from the same problems that beset many of the scarce metals—they are geographically restricted.

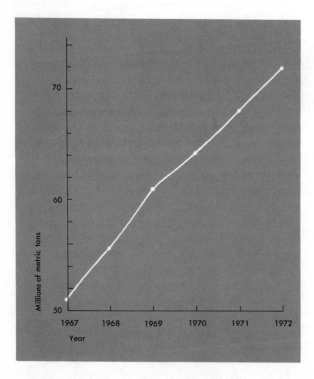

FIG. 7-1 World consumption of phosphorous, potassium, and nitrogen, the principal fertilizer elements. The fertilizer elements are applied to the soil as compounds, making the total tonnage of fertilizer materials much greater than shown on the graph. The weight ratios of the fertilizer elements are approximately 1 to 1.5 to 3 for phosphorous, potassium, and nitrogen, respectively. (Data from U.S. Bureau of Mines.)

Nitrogen

As mentioned in Chapter 2, nitrogen is the principal resource won from the atmosphere. The form in which it is added to the soil is either as a soluble nitrate, such as KNO_3 or $NaNO_3$, or as an ammonia compound, such as $(NH_4)_2SO_4$ or NH_4NO_3.

Most synthetic nitrogen compounds are produced either by one of the variants of the Haber-Bosch process—in which nitrogen from the atmosphere and hydrogen are combined under high temperatures and pressures to form ammonia—or as by-products from coke ovens. The process uses a great deal of energy and nitrogen is the largest contributor to the fertilizer curve in Fig. 7-1. The cost of fertilizers is strongly influenced, therefore, by the cost of energy. Despite the increasing dominance of atmospheric nitrogen as a source of nitrogenous compounds, a small production of natural nitrates from extensive Chilean deposits continues. Although nitrates are very soluble compounds, they can accumulate in the soils of Chile's Atacama Desert because it has the world's lowest rainfall. These are now mined and yield about 125,000 metric tons of nitrogen a year. Natural nitrates are not, however, essential mineral resources, the synthetics being more than adequate as replacements when necessary.

Almost one-quarter of the nitrogen consumed annually is produced in the United States—10,350,000 metric tons compared with a world total of 41,140,000 in 1971.

Potassium

Potassium is an abundant element, widely distributed in silicate minerals. It is the rarely formed soluble minerals that we seek as resources, however, and these are almost solely confined to a special class of mineral deposits, known as *marine evaporites*, that result from the accumulation of salts by evaporation of sea water. Because marine evaporites are also of vital importance in the production of other minerals, we will briefly discuss their origin.

The major elements dissolved in sea water were displayed in Fig. 2-3. They can be recast into the constituents that actually precipitate from sea water by balancing the positively charged cations, such as sodium (Na^{+1}), against the negatively charged anions, such as chlorine (Cl^{-1}), to preserve electrical neutrality (Table 7-1). Sodium chloride is the most abundant constituent; next follow the magnesium salts, then calcium sulfate, and potassium chloride. When water is removed by evaporation, brine becomes increasingly concentrated and finally reaches saturation, first in one salt, then in the others. It is not necessarily the most abundant compound that precipitates first; saturation of the relatively insoluble—and therefore sparse—compounds is usually reached long before saturation of those that are highly soluble.

Table 7-1 Major Constituents of Sea Water*

Constituent	Percentage of Total Dissolved Solids
NaCl	78.04
$MgCl_2$	9.21
$MgSO_4$	6.53
$CaSO_4$	3.48
KCl	2.11
$CaCO_3$	0.33
$MgBr_2$	0.25
$SrSO_4$	0.05

* Obtained by recasting the data in Fig. 2-4 into molecular compositions.

The first compound to precipitate from evaporating sea water is $CaCO_3$, for which the solubility is extremely low and the amount in solution small relative to NaCl. The next compound, $CaSO_4$,* does not separate until the solution has been reduced to 19 percent of the original volume, and NaCl, the third to separate, only does so when the residual solution is reduced to 9.5 percent. Precipitation of NaCl, plus a small amount of $CaSO_4$, continues until the brine is reduced to about 4 percent of its original volume; then the first compound to contain either magnesium or potassium, a complex salt called polyhalite ($K_2SO_4 \cdot MgSO_4 \cdot 2CaSO_4 \cdot 2H_2O$), begins to precipitate. The amount of NaCl in solution is large to begin with, and considerably more than half of it will be precipitated during the reduction in solution volume from 9.5 percent to 4 percent, so the thickest layer formed during a single evaporation cycle will be the NaCl layer. The sequence of minerals separating from the final 4 percent of the brine (called the bitterns) is complex and variable, depending on such factors as the temperature and whether or not the final liquid remains in contact, and hence can react with the earlier-formed crystals. Two of the precipitates found in most sequences are sylvite (KCl) and carnallite (KCl $\cdot$ $MgCl_2$ $\cdot$ $6H_2O$), and it is in these late-stage evaporite minerals that most of the world's useful resources of potassium minerals are to be found.

The evaporation of a completely isolated body of sea water would produce the sequence and volume of salts shown in Fig. 7-2. When we examine actual evaporite deposits, these volume relationships are rarely found—the early-formed precipitates, $CaCO_3$ and $CaSO_4$, are much more abundant, and the

---

* Either $CaSO_4$ or $CaSO_4 \cdot 2H_2O$ may precipitate depending on the temperature. Discussion of these two compounds can be found in the section on plaster in Chapter 8.

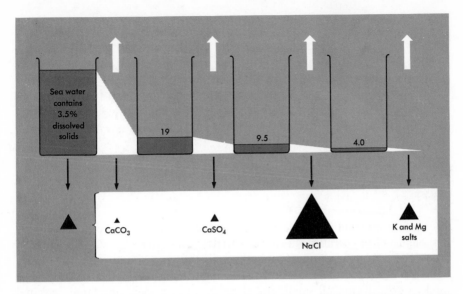

FIG. 7-2 Succession of compounds precipitating from sea water. When evaporation reduces the starting volume to 19 percent, CaSO₄ begins to precipitate; at 9.5 percent, NaCl; and so on.

late-formed precipitates, the K and Mg salts, are rare. Furthermore, complete evaporation of a body of sea water even as deep as the Mediterranean, which averages about 1,370 meters, would produce only 24 meters of NaCl and a layer of $CaSO_4$ only 1.4 meters thick. However, beds of $CaSO_4$ and NaCl several hundreds of meters thick are known geologically and from evidence such as fossils in associated rocks it can be deduced that evaporation occurred in shallow bodies of water. Clearly, some mechanism other than the drying up of an isolated, shallow basin must occur. The common circumstance is to find precipitation in a partially isolated basin from which water is removed by evaporation but into which fresh sea water is continually flowing. There are several geologic circumstances in which such an arrangement occurs, and they are collectively called *barred basins*. Water flows into the basin over a submerged bar; evaporation of the surface waters continually enriches the basin in dissolved salts because the partially enriched but heavier brine sinks to the bottom and is prevented from recirculating by the restricting bar (Fig. 7-3). The salinity of the basin increases more slowly than it would by direct evaporation of an isolated body, and the brine remains for a long time in the salinity range in which $CaCO_3$ precipitates. It is even possible for the precipitated $CaCO_3$ completely to fill the basin to the level of the bar before a sufficiently high salinity of the precipitation of $CaSO_4$ is reached. Similarly, thick beds of $CaSO_4$ may form and the brines never reach the NaCl stage. The frequency of occurrence *and* the total thickness of evaporite salts in sedimentary basins around the world *decrease* in the order $CaCO_3$ > $CaSO_4$ > NaCl > K and Mg salts.

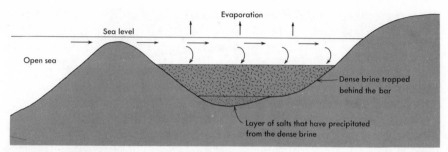

FIG. 7-3 Diagrammatic cross section of a barred basin. Fresh sea water flows over the bar and is concentrated by evaporation. The dense brine sinks and is prevented from returning to the open sea by the bar. When the brine reaches a sufficiently high salinity, salts precipitate and may eventually fill the basin.

At the present time evaporite deposits are accumulating on every continent. Although high temperatures are not essential for concentration—drying winds can accomplish the same thing in cold climates—the evaporites are formed mostly within a belt 35 degrees north latitude and 35 degrees south latitude. The modern deposits are all small, and there is not at present any large marginal sea with a restricting flow—such as the Baltic, the Mediterranean, or the Black Sea—that fulfills both the morphological and the necessary climatic conditions of a large evaporite deposit. As we look at the geological record, however, we find that this is a temporary situation, for marine evaporites are widely spread both in time and in space, and there have been several times in the past when worldwide conditions, such as prolonged periods of higher temperatures, were much more conducive to the formation of marine evaporites than they are at present. Thus, during the Permian period exceptionally thick evaporite sequences were formed in North America and Europe. Indeed, the word Permian is taken from the Perm Basin in the U.S.S.R. where one of the world's largest evaporite deposits is known.

Evaporites are our principal source of halite (NaCl, familiar to us as common salt), of gypsum ($CaSO_4 \cdot 2H_2O$), and of potassium salts. In some cases magnesium and less abundant elements are produced as by-products. Halite and gypsum will be discussed later. For the present we will concentrate on potassium compounds.

For many years most of the world relied on the vast Permian evaporite deposits in Germany. It is somewhat ironic that the circumstances of World War I, which closed the German deposits to the United States, should have been responsible for providing the impetus to geologists to find even richer deposits of the same type and age in New Mexico. A large, shallow Permian sea deposited a thick section of evaporite salts over an area of at least 160,000 square kilometers in what is now New Mexico, Texas, Oklahoma, and Kansas. In a portion of the basin, near Carlsbad, New Mexico, an estimated 4,800 square kilometers of the sequence contain potassium salts in beds that reach 4 meters in thickness. These deposits, which are among the richest in the world but small by comparison with several others, have indicated reserves of potassium of about 100 million metric tons.

North America has other large potassium reserves (Fig. 7-4, Table 7-2). A basin of Pennsylvanian age in southeastern Utah and southwestern Colorado,

FIG. 7-4 Areas of the United States and Canada underlain by major marine evaporite deposits of halite and potassium salts. (After *U.S. Geol. Surv. Bull. 1019-J*; A. D. Huffman, 1968; P. B. King, 1942; R. J. Hite, 1961; and S. R. L. Handing and H. A. Gorrell, 1967.)

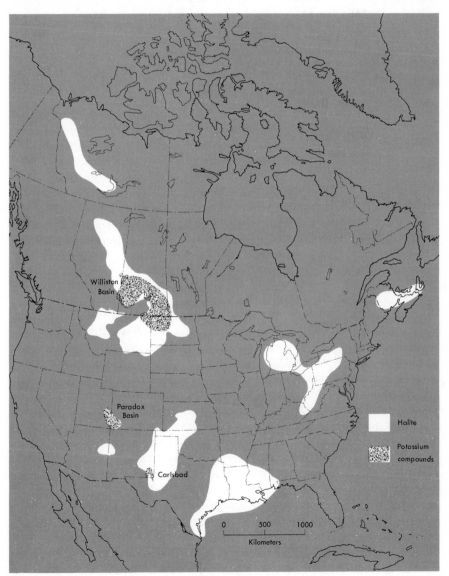

Table 7-2 Production and Reserves of Potassium, 1971*

Country	Production (metric tons)	Reserves (metric tons)
U.S.S.R.	4.44×10^6	$37{,}640 \times 10^6$
CANADA	2.91×10^6	$35{,}800 \times 10^6$
WEST GERMANY	2.42×10^6	$7{,}200 \times 10^6$
EAST GERMANY	2.03×10^6	$7{,}500 \times 10^6$
U.S.A.	1.95×10^6	340×10^6
WORLD TOTAL	16.66×10^6	$99{,}120 \times 10^6$

* Because several different potassium compounds are mined, it is convenient to express production in terms of the element potassium.

(After U.S. Bureau of Mines.)

called the Paradox Basin, contains an estimated 12,600 square kilometers of potassium-rich salines, though much of it is too deep to warrant present recovery. In Saskatchewan, Canada, a huge and as yet incompletely explored resource of potassium salts has been found in the Williston Basin (Devonian period). Estimates of as much as 35,800 million metric tons of potassium that are accessible by today's mining standards have been published. The largest known reserves are in the U.S.S.R., however, and ultimately the salts dissolved in the sea present a nearly inexhaustible potential resource. Even great expansions beyond today's production rates would not seriously threaten potassium supplies.

Phosphorus

The phosphorus cycle appears to be the one most disrupted by intensive land cultivation. Phosphorus is a vital element for many cellular processes in the body, but man's major store of it is in his skeleton, composed of the mineral *apatite* ($Ca_5(PO_4)_3OH$), and this supply is not returned to the soils of our crop lands after death; it is therefore necessary to return it as a fertilizer. The only important source of phosphorus, and the only common phosphate mineral, is apatite; however, it is relatively insoluble. It is common practice, therefore, to treat apatite (or *phosphate rock*, as most of the apatite-rich raw materials are called) with dilute acids, usually sulfuric (H_2SO_4), to produce a more soluble material. The chemical reactions that ensue are complex, but the result, known as *superphosphate*, contains a high percentage of water-soluble compounds such as $Ca(H_2PO_4)_2$.

Apatite is widespread in trace amounts in most rocks, regardless of whether they are igneous, metamorphic, or sedimentary. Major concentrations of apatite in igneous rocks are known, but they are comparatively rare; one such concentration, on the Kola Peninsula in the northern Soviet Union, is mined for its apatite content and is a major supplier of Soviet phosphatic fertilizers. The major production of apatite, however, accounting for more than 90 percent of the present world production and for almost all of our known reserves, is from marine sedimentary deposits.

The origin of the marine deposits is still in doubt; as with other geologic problems, we have not been able to solve it by observing similar beds forming today. In general outline, the process is believed to arise in the following manner. The waters of the ocean are near saturation with respect to apatite, and several effects can apparently cause saturation to be exceeded and precipitation to ensue. For example, nearly oxygen-deficient, or anoxic, waters provide environments in which the water composition is considerably less alkaline than normal sea water and in which the solubility of apatite is exceeded. These environments are not common in the sea, especially on a large scale, and when they do form, the basin must be so situated that very little detritus is introduced to dilute the slowly precipitating apatitic sediments.

A particularly large and apparently unique deposit of this type was formed during Permian times in a shallow marine basin covering what are now portions of the states of Idaho, Nevada, Utah, Colorado, Wyoming, and Montana in the United States (Fig. 7-5). The phosphate-rich sediments, called the Phosphoria Formation, cover more than 160,000 square kilometers and reach thicknesses of 140 meters. Over most of the area the thickness of the phosphatic bed is only

FIG. 7-5 Extensive phosphate deposits are worked in three important areas of the United States. The largest reserves are in the Phosphoria Formation and appear to be adequate for many centuries, provided that mining and transportation problems can be solved. (After V. E. McKelvey et al., 1953; Tenn. Div. of Mines, 1938; *U.S. Geol. Surv. Bull.*, 1942.)

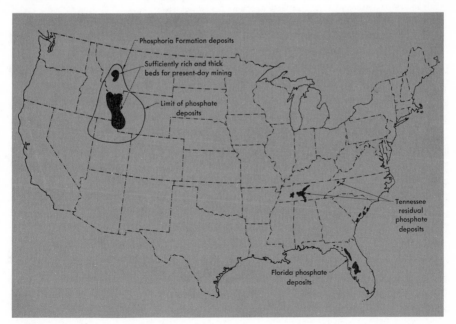

a meter or less, and at best it can be considered only a potential resource. The tonnage, however, is enormous.

Small nodular bodies of precipitated apatite are common in some limestones formed in shallow marine water. Large concentrations of nodules, whether formed directly or concentrated secondarily into gravel beds, form valuable phosphate deposits. The huge "land pebble" phosphate deposits of Florida, source of most of the present phosphate production in the United States, are secondary concentrations of phosphatic nodules from Miocene limestones. Another large producing area, in Tennessee, contains residual deposits of phosphates formed by the weathering of nodule-containing limestones.

The principal phosphate-producing countries of the world are also the principal reserve-holders, and the reserves are indeed huge (Table 7-3). The

Table 7-3 Production and Reserves of Phosphorus, 1971*

Country	Production (metric tons)	Reserves (metric tons)
U.S.A.	4.2×10^6	$6,200 \times 10^6$
U.S.S.R.	2.6×10^6	$2,400 \times 10^6$
MOROCCO	1.4×10^6	$8,300 \times 10^6$
TUNISIA	0.4×10^6	500×10^6
WORLD TOTAL	10.4×10^6	$19,800 \times 10^6$

* Phosphorus production is reported either as tonnage of phosphate rock mined, or as P_2O_5. Average phosphorus contents of the different ores have been used to construct the table.

(After U.S. Bureau of Mines.)

spotty geographic location of known resources is not encouraging, however, for transportation costs add heavily to the expense of the fertilizer. Countries such as Australia, with large land areas needing phosphatic fertilizers, have therefore actively pursued local exploration programs in the hope of finding accessible supplies. In the case of Australia, a large deposit has been recently located in western Queensland, but, unfortunately, the transportation costs from this remote location to the agricultural parts of the country have delayed exploitation.

Potential resources of phosphorus are large. Prospecting for phosphate deposits has not been thorough enough to be certain that even very large deposits may still be undiscovered. In part this is a recognition of the economic difficulties entailed in opening new deposits in competition to existing mines. But in part, too, it stems from the difficulty of recognizing a phosphorus-rich rock. Even to an expert, many phosphate rocks look like ordinary shales and limestones. We can, therefore, probably anticipate discoveries of new, large deposits in the future.

One large potential resource has already been discovered. Along many continental shelves of the world there are crusts and nodules of apatite formed during the Miocene—these are the same formations that yield the rich phosphates in Florida and Morocco. Unfortunately, the deposits are lower grade than their landward equivalents, but the tonnages are large, probably as large as the present reserves. We must therefore conclude that phosphorus, like nitrogen and potassium, will not soon become a limitation to food production.

Sulfur

Because of its diverse uses, sulfur is not commonly thought of as a fertilizer element; approximately 40 percent of the world's production, however, is used in the manufacture of superphosphate and ammonium sulfate, both essential fertilizers. The second-largest consumer, the chemical industry, takes 20 percent and uses much of this for the production of insecticides and fungicides for crop protection.

Since sulfur is an abundant element, its sources are varied and widespread. The sea, for example, contains vast resources of *sulfates*, and the huge evaporite deposits of the world, discussed previously, contain enormous resources of $CaSO_4$. However, sulfur, like other resources, is preferentially sought in its least costly supply, this being the *native* form.

There are only two important resources of native sulfur. One, exploited principally in Japan, is from volcanoes, which give off sulfurous gases that condense in near-surface veins and rock impregnations. The other source, quantitatively much larger, is derived by secondary concentration from $CaSO_4$. Certain anaerobic bacteria derive their oxygen from solid compounds, such as $CaSO_4$, and their food supplies from decayed organic matter. A series of reactions ensues, by which the bacteria change the $CaSO_4$ to $CaCO_3$ and S.

The places where this process occurs are those where petroleum is present to supply the bacterial food. In the United States and Mexico this has occurred in two environments: on top of salt domes and in certain evaporite beds. The origin of salt domes will be discussed later; for the present it is sufficient to state that the source of the salt is deeply buried marine evaporites, and that the associated $CaSO_4$ is brought to a near-surface environment where the bacterial reduction proceeds. Although several hundred salt domes have been found around the world, only a few contain commercial quantities of native sulfur. These occur along the coast of the Gulf of Mexico from Alabama to Mexico; they are very rich, however, and account for 25 percent of the world's current production. The second source is much larger than the salt dome source, but because it has only recently been recognized, it is still a relatively small contributor to the present production. If circulating subsurface waters should locally dissolve $CaSO_4$ from an evaporite bed, petroleum and bacteria can enter the resulting voids and thus lead to the formation of rich localized zones of sulfur. This has happened on a large scale in west Texas, where more than 50×10^6

metric tons of recoverable sulfur have been found in the Culbertson Field alone.

Compared to consumption, the reserves of native sulfur are not large, and since World War II a determined drive on the part of many countries has opened two alternative sulfur sources. The first is the H_2S, or *sour gas*, content of natural gas. Formerly allowed to escape, the H_2S component is now widely recovered and oxidized to sulfur. The second source is in the sulfide ores of the scarce metals, recovered as a by-product, and in deposits of two iron sulfides, pyrite (FeS_2), and pyrrhotite (FeS).

The world production of sulfur from all sources reached 29,000,000 metric tons in 1971; of this, 10,600,000 metric tons were produced by the United States. Reserves of rich native sulfur ores are relatively small, but resources of sulfide and sulfate ores are huge. Provided that technological development can keep the price of sulfur low as the alternative resources are used, we shall always have abundant supplies.

MINERALS FOR CHEMICALS

The nonmetallic minerals used principally for raw materials in the chemical industry are diverse and have considerable economic importance. They have few claims to being essential compounds, however, and the use of most is guided chiefly by their great abundance, easy recovery, and hence low cost. Substitutes are already available for many.

The most important member of the group is NaCl. Not only does the sea contain vast resources of salt (see Chapter 2), but the marine evaporites also contain such vast deposits that the problems of man's resources essentially become problems of the practicality of mining and shipping (Fig. 7-4). The previously mentioned Permian evaporite basin of the south and central United States, for example, contains more than 150,000 square kilometers of halite beds aggregating 60 meters in thickness, but the present cost of mining the deposits is high because of their depth below the surface as well as their distance from major markets; they are exploited only in one small area in Kansas.

Although many beds are too deep to mine, nature has an interesting way of bringing some salt nearer the surface. Halite has a density of 2.2 grams per cubic centimeter, whereas most of the associated sedimentary rocks have densities of at least 2.5 grams per cubic centimeter. The salt beds, being lighter and capable of plastic flow like ice in a glacier, tend to rise and "flow" up through the overlying rocks. Provided that the overlying rocks are still weak enough to be ruptured by the rising salt, long thin columns or plugs of salt will float their way up from deeply buried sedimentary salt horizons. Columns ranging from about a hundred meters to more than two kilometers in diameter are known to have risen up through as much as twelve kilometers of overlying sediments.

Salt domes are known in many areas of the world—Europe, South America, the Middle East, and the U.S.S.R.—but they are particularly frequent in the

FIG. 7-6 Known and probable salt domes (white dots and open circles, respectively) that have been identified in the coastal area of the Gulf of Mexico. In no other part of the world have so many of these unusual bodies been identified. [Adaptation of Fig. 5.1, p. 203, *Geology of the Atlantic and Gulf Coastal Province of North America* by Grover E. Murray (Harper and Row, 1961).]

area bordering the Gulf of Mexico, where several hundred have been identified (Fig. 7-6). These domes have risen from evaporite beds 12,000 meters below the present flat coastal plain of Louisiana and Texas, and several are mined for salt.

Salt is an essential ingredient in our diets, and 99 countries produce it on a regular basis. Direct human consumption is only a small portion of the total, however, since most goes for the manufacture of chlorine and soda ash (Na_2CO_3) in the chemical industry and for road control of ice and snow. Not surprisingly, the major industrial powers—the United States, the U.S.S.R., Germany, and the United Kingdom—consumed 50 percent of the 1971 production of 142,700,000 metric tons.

Of the raw chemical group, with the exclusion of petrochemicals derived from fossil fuels, salt has the largest production. Others of importance are Na_2CO_3, used for production of paper, soap, and detergents, and for water treatment; Na_2SO_4, used for kraft paper, detergents, and additives in tanning and dyeing; and borate minerals, such as borax ($Na_2B_4O_7 \cdot 10\ H_2O$), used for glass-making fluxes, soaps, detergents, hide-curing compounds, and antiseptics. These deposits are all formed by precipitation from lake waters in *nonmarine evaporite deposits* and are available in large quantities.

OTHER INDUSTRIAL MINERALS

The list is large but only one group, the abrasive minerals, is so technologically essential that it requires discussion.

Abrasives are becoming increasingly important for working hard modern alloys and compounds such as tungsten carbide, and it is diamond, the hardest known natural compound, that is the most important resource. Diamond—the most dense natural form of carbon—requires high pressures for its formation; these are reached only at depths of 150 kilometers or more in the Earth. The diamond-bearing rocks from these great depths, called kimberlites, come from the mantle and are themselves rare. They reach the surface in narrow pipe-like vents, often no more than 50 meters in diameter, and the reasons for the formation and location of the pipes remain a geological puzzle.

Kimberlites are the home of the diamond, but not all kimberlites contain diamonds. Of the several hundred kimberlite pipes found in Africa to date, only 29 are known to contain diamond, and of these 12 are too lean to warrant mining. On other continents the percentage is even lower, and, outside Africa, only in the Yakutia area of the U.S.S.R. have kimberlites been discovered that are rich enough to be worked. Even the richest kimberlites contain very low diamond contents—no more than 0.0000073 percent.*

Because diamonds are dense and almost indestructible, they accumulate in placer deposits; 96 percent by weight of all diamonds recovered still comes from placers. Although we immediately think of gems when diamonds are mentioned, only 20 percent of the diamonds produced can be so cut—though this still amounts to 65 percent of the monetary value of diamond production—and the remaining 80 percent is used for such industrial purposes as cutting, die-making, and the manufacture of abrasives.

From the time of their discovery in 1870, African deposits have been the world's largest producers of diamonds—first in South Africa, with kimberlite pipe production unusually rich in gem material, but more recently in Zaire and in Ghana—and these have produced over 80 percent of the world's diamonds (Table 7-4). Recent finds in northern Siberia account for a currently growing

Table 7-4 Production and Reserves of Industrial Diamonds, 1971

Country	Production (grams)	Reserves (grams)
ZAIRE	2.74×10^6	100×10^6
U.S.S.R.	1.76×10^6	5×10^6
SOUTH AFRICA	1.4×10^6	11×10^6
GHANA	0.5×10^6	5×10^6
ANGOLA	0.4×10^6	1×10^6
WORLD TOTAL	8.4×10^6	126×10^6

* This is equivalent to 0.2 gram (1 carat) for each three metric tons of rock.

Soviet production. Measured reserves of diamonds largely in Zaire are reported to be in excess of 100,000,000 grams, but clearly they are not large enough to satisfy demands for long.

As it has with other abrasives, technological progress has finally shown the way to self-sufficiency in diamonds. In 1955 the General Electric Company announced its successful synthesis of industrial diamonds, using a special ultra-high-pressure reaction vessel. Commercial production has grown to a presently estimated level of 2,000,000 grams per year, with production coming from Sweden, South Africa, Ireland, Japan, and the Soviet Union, in addition to the United States.

eight

building materials

There will be a shortage of standing room on earth before there is a shortage of granite. (J. A. S. Adams, New Ways of Finding Minerals, 1959.)

Building materials are the largest crop and, after fossil fuels, the second most valuable mineral commodity that we reap from the Earth. Because almost every known rock type and mineral contributes to the crop in some way, the origin of building materials embraces most of geology. We will concentrate mainly on uses and try to place this largest-volume mineral production into perspective with the other mineral products.

Although no classification is completely satisfactory, we will separate building materials into two groups: first, materials that are used as they come from the ground, without any treatment beyond physical shaping, such as cutting or crushing; second, the prepared materials that must be treated chemically, fired, melted, or otherwise altered before use so that they can be molded and set into new forms. The first group includes building stones, sand, gravel, and crushed stone for aggregate; the second includes clay for bricks, raw materials for cement, plaster, and asbestos.

Most building materials, unlike metals, have little intrinsic value; they are not scarce commodities and they are widely distributed, but when removed and processed to a useful form, they increase in value enormously. For example, the limestone and shale used to make cement may have intrinsic values of only $1 per ton or less in the ground, but after mining, crushing, firing, and conversion to a high-quality cement, the product is worth $20 or more per ton (Fig. 8-1). The factors controlling location of production sites are most commonly such straightforward ones as local demand and transport costs; rarely are they problems of resources.

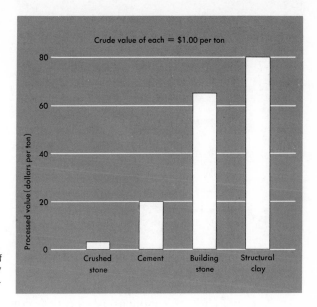

Crude value of each = $1.00 per ton

Processed value (dollars per ton)

80

60

40

20

0

Crushed stone   Cement   Building stone   Structural clay

FIG. 8-1 Processing of building materials adds greatly to their value. (After W. L. Fisher, 1965.)

NATURAL ROCK PRODUCTS

There are three important classes of natural rock products, and the supplies of each are almost limitless. It makes little sense to consider global reserves, therefore, since production problems derive from costs of transportation rather than from abundance.

Building Stone

The uses of building stones range from roofing slate and curbstones to facings for public buildings and tombstones. Though used from time immemorial as structural and foundation materials, as in the Egyptian pyramids, building stone is being supplanted by concrete; its remaining uses largely are governed by the pleasing ornamental display of natural stone.

Figures on production volume and types of stone used are not available on a worldwide basis; however, the best-documented production—that from the United States—probably gives a reasonable idea of the global balance. The total tonnage of building stones mined in the United States is about 2,000,000 metric tons. The two most popular rock types are limestone and granite, jointly aggregating more than 60 percent of the total (Fig. 8-2).

What are the problems with building stones? When stones of pleasing appearance and good physical properties have been found, problems mostly are to do with mining: how to mine the rock without shattering it—which means that extensive blasting is not possible—and how to select areas where natural joints and cracks in rocks are a help, rather than a hindrance, to mining.

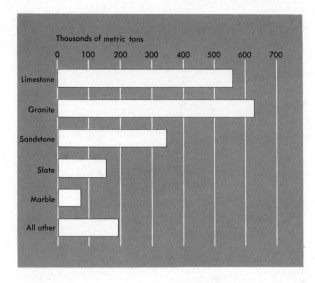

FIG. 8-2 Types of building stones used in the United States. Although direct figures are unavailable, it is probable that the same approximate range of rock types is used throughout the world. (After U.S. Bureau of Mines.)

Crushed Rock

A commodity of enormous proportions, crushed rock accounted for 800 million metric tons in 1972 in the United States alone. Yet little more than a century has passed since Eli Whitney Blake, responding to the call for quantities of crushed rock needed for the then ambitious project of a two-mile long macadam road from New Haven to Westville, Connecticut, invented the modern rock-crusher in 1858. Before this, all rock crushing was done by hand.

Crushed stone is still used principally for roadbeds and for concrete aggregate, although about 15 percent, principally limestone, is used as raw material for the manufacture of cement. The most widely used rock types are limestone and dolomite, both easy to mine and crush, but strong in use. Basalt and other fine-grained, dark-colored igneous rocks, commonly called *trap rock* by industry, are a poor second (Fig. 8-3). The only essential requirement for crushed stone is a rock outcrop to quarry, and indeed rockcrushers can be seen—and heard—adjacent to most cities in developed countries.

Sand and Gravel

Sand and gravel, largely used like crushed rock for highway roadbeds and for concrete aggregate, are consumed in amounts even larger than those of crushed stone. The consumption in the United States in 1972 was 835 million metric tons. Sand is classified as having particle sizes less than 2 millimeters in diameter; those of gravel are larger. The tonnage of gravel used is twice that of sand.

The geological domain of gravel is that of streams in which the rounded pebbles are produced by continuous movement of fast water and in which a size separation occurs on the basis of weight: finer grains are washed downstream or out to sea. Local resources of sand and gravel may often be limited, especially in flat regions and places devoid of large rivers. Where land resources are being depleted too rapidly, exploration and exploitation of offshore marine sandbars may also be employed. An interesting situation has now been reached in both Europe and North America. Population densities in coastal areas above about 40 degrees north latitude have very nearly consumed all the sand and gravel. It is precisely in these high latitudes, however, that extensive sand and gravel deposits occur on the continental margins, having been deposited there by glaciers during the height of the recent ice age. These communities therefore have alternatives; either quarry and crush rock or dredge the sand and gravel from beneath the sea. Off the west coast of Europe approximately $100 million worth of sand and gravel are now dredged each year. In North America dredging has commenced in a small way, and it can be expected to grow in the years ahead, particularly off the shores of New Jersey, New York, and the New England states.

Gravel deposits are sparse or absent off many tropical coasts. Around the Gulf of Mexico, for example, sands can be found in river deltas, but gravels are nearly unknown. In the Gulf coast area igneous and metamorphic rock do not reach the surface, so crushed rock cannot be made. The only coarse building material under the circumstances are old shell beds, particularly oyster shells, clam shells, and coral reefs. As these limited supplies are used up, areas such as southern Texas, Louisiana, and Florida will be forced to bring in gravel or crushed stone from elsewhere.

PREPARED ROCK PRODUCTS

From the day man first molded a clay object and fired it in his hearth, he has been using prepared rock products. Although we now process and use a bewilderingly large array of materials, five of these account for most of the volume and value.

FIG. 8-3 Rock types most widely used in the preparation of crushed stone in the United States. (After U.S. Bureau of Mines.)

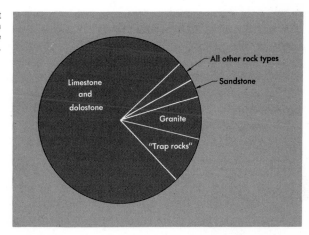

Cement

The word cement refers to agents that bind particles together. Cement is rarely used alone; it is added to sand, gravel, crushed rock, oyster shells, or other aggregate as the binder needed to make concrete, a sort of "instant rock." As little as 15 percent of a concrete may be cement.

The forerunner of modern cement was discovered by the engineers of ancient Rome. They found that water added to a mixture of quicklime (CaO, obtained by heating—or calcining—limestone) and a natural, glassy volcanic ash from the town of Pozzuoli, near Naples, produced a series of reactions that caused the mixture to recrystallize and harden. The resulting mass was stable in air or water and it materially assisted the Romans in their remarkable engineering feats. Known as *pozzolan cement*, similar materials are still employed today, but on a decreasing scale. In a pozzolan cement one of the ingredients (quicklime in ancient Rome) is calcined to get it in a reactive form; the others are naturally reactive materials. However, the volcanic ash had also been essentially calcined by nature during the volcanic eruption; why should all the ingredients not be calcined by man if the right rock compositions were selected? Apparently the Romans realized this too, for they discovered the secret of cement manufacture.

The "secret" was forgotten during the Dark Ages, and it was only redis-covered in 1756. John Smeaton, a British engineer engaged in designing and building the famous Eddystone Lighthouse, sought cementing materials to set and remain stable under water. He is said to have rediscovered the Roman cement formula through examination of an ancient Latin document. Natural cements soon became popular in Europe and subject to much experiment. In 1824 another Englishman, Joseph Aspidin, patented his formula for *portland cement*, so called because it resembled Portland stone, a limestone widely used in British buildings. It soon supplanted all other cements and is today the most common construction material in the world, with twice as much concrete being used as all other structural materials combined.

Production of portland cement has grown erratically, but continuously, at a highly exponential rate as population and technological expansion both increase; and the growth is expected to continue. Not surprisingly, the largest producers and consumers are the highly industrialized countries (Table 8-1).

Portland cements have a range of composition depending on their use and are prepared by heating ground rock of suitable composition to a temper-ature of approximately 1,480°C. This expels all the carbon dioxide and water and causes part of the charge to melt to a glass. The resulting clinker is crushed and is then ready for use; when it is mixed with water, a series of chemical reactions proceed in which new compounds form and grow as a hard cemented mass of interlocking crystals.

The raw materials for portland cement (Fig. 8-4) are found in a suitable mixture of a slightly dolomitic limestone and a shale or clay. Since limestone is the largest ingredient, cement works are usually situated close to a suitable

Table 8-1   Producers of Cement, 1971*

Country	Production (metric tons)	Percentage of Total
U.S.S.R.	100.3×10^6	17.0
U.S.A.	72.8×10^6	12.3
JAPAN	53.7×10^6	9.1
WEST GERMANY	32.7×10^6	5.5
ITALY	31.7×10^6	5.4
FRANCE	28.9×10^6	4.9
WORLD TOTAL	590.1×10^6	

* 110 different countries produced 20,000 metric tons or more cement in 1971, but the six large industrial countries listed in the table accounted for 54.2 percent of the total production.

(After U.S. Bureau of Mines.)

source and other ingredients are transported in. The most desirable circumstance is a somewhat impure limestone in which the impurities are clays of the desired composition. Such beds are indeed found, and natural cement rocks now account for approximately 20 percent of the cement production.

Plaster

Plaster, made by heating or calcinating gypsum ($CaSO_4 \cdot 2H_2O$), is one of man's oldest building materials, and the consumption continues to grow,

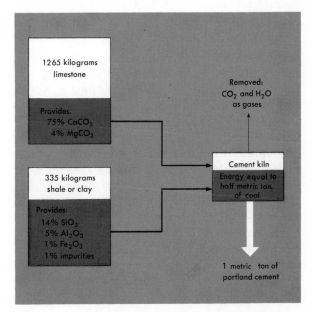

FIG. 8-4 Typical mix of raw materials needed to produce a widely used composition of portland cement.

1265 kilograms limestone

Provides:
75% $CaCO_3$
4% $MgCO_3$

335 kilograms shale or clay

Provides:
14% SiO_2
5% Al_2O_3
1% Fe_2O_3
1% impurities

Removed:
CO_2 and H_2O
as gases

Cement kiln

Energy equal to half metric ton of coal

1 metric ton of portland cement

with a worldwide production of 54 million metric tons in 1972. Calcining gypsum at 177°C rapidly drives off 75 percent of the water and changes it to a new compound, $2CaSO_4 \cdot H_2O$, commonly called *plaster of Paris*, after the famous gypsum quarries in the northern part of that city, from which a plaster of particularly high quality is produced. When plaster of Paris is mixed with water, it reverts to a finely interlocked mass of tiny gypsum crystals. Plaster can be used straight, mixed with sand, or mounted on wallboard or paper backings for prefabricated finished surfaces.

In Chapter 7 the precipitation of $CaSO_4$ in evaporite sequences was discussed. The precipitate may take one of two forms, either gypsum or anhydrite ($CaSO_4$), so named because it does not contain any water of crystallization. Which of the two forms precipitates is a function of temperature, with anhydrite being favored by higher temperatures; under the hot climatic conditions where most evaporites form, anhydrite is the common precipitating form. It is gypsum that we usually find today, however, for the ground temperature in the shallow deposits where it is worked is low, and with water percolating down from rainfall, the anhydrite soon hydrates to gypsum in the same way that plaster of Paris does.

Gypsum and anhydrite are widely distributed; more common than salt but less common than limestone, they are known throughout the geologic record. A resource map published by the U.S. Geological Survey shows that as much as 10 percent of the land area in the United States is underlain by gypsiferous rocks, which indicates that potential resources are so large as to be superabundant. Indeed, the U.S. Bureau of Mines has stated that reserves in the United States alone are sufficient for 2,000 years at projected rates of production.

Clays

There are many structural and refractory ceramic materials now used and discussion of each would be pointless. Most are formed from clay that can be molded into desired shapes, then fired to hardness. Bricks so formed have been used at least since the days of the Babylonian Empire, first in a form with low durability—the bricks having been allowed to dry in the sun—but later in forms of increasing durability, achieved by firing and even by glaze-coating. As with cement, the art of brickmaking and clay work reached a high degree of perfection in the days of the Roman Empire, but much of the art was lost in Europe during the Dark Ages. From the time of the industry's revival in Europe in the thirteenth century, production of clay for structural ceramics has grown steadily—first for the preparation of bricks, but later for tiles, drainpipes, and numerous other uses. By 1972 the production of clay in the United States alone was 52 million metric tons, and of this, 40 percent was for structural ceramics, such as bricks, sewer pipes, and tiles, and 20 percent for production of cement. The remainder finds a great variety of uses, but the principal ones are the manufacture of light-weight concrete aggregate; fillers for paper, plastics, rubber, and paint; absorbent uses such as for animal litter and oil cleaning; oil well drilling muds, and pottery.

Clays are formed by weathering at the Earth's surface and may accumulate as residual deposits, as has been discussed in "Aluminum" (Chapter 5); they also may be transported and deposited as sedimentary clay beds. Whether transported or residual, clays soon begin to lithify and become solid rocks when they are dehydrated and heated by burial. Their environment is the Earth's surface and not the deeper portions of the crust. Like the other building resource materials, therefore, they are usually recovered by large quarrying operations in surficial deposits, with the quarries preferentially situated as close to major consuming sites as possible.

Reserves of clays are so large that few countries make attempts to estimate them. Those of the United States, for example, are said by the U.S. Bureau of Mines to be more than sufficient for another century of production at anticipated rates of growth. Potential resources are even larger.

Glass

We might not immediately think of glass as belonging in a category with bricks and cement, but it consumes similar raw materials and is processed in a similar fashion. Futhermore, glass has begun to challenge older, more established structural materials in many specialized uses, and consumption is rising rapidly.

Glass is made by melting rocks and minerals, then quenching them so rapidly that crystals do not have time to nucleate. This procedure is more readily carried out with some materials than with others, and most readily with silica (SiO_2), usually obtained from quartz in sandstones. The melting point of quartz is very high ($1,713°C$), and to reduce this to a more easily reached temperature, ingredients such as CaO (from limestone), Na_2O (from sodium carbonate), and borax are added.

Asbestos

Asbestos is the name given to the fibrous form of many minerals. One such, chrysotile ($Mg_3Si_2O_5(OH)_4$), a fibrous form of the serpentine minerals, is so abundant and has such excellent physical properties that it accounts for more than 90 percent of all the asbestos minerals mined. Asbestos fibers, being strong and flexible, can be spun and woven like organic fibers such as cotton and wool. The resulting products are not flammable, are good electrical and thermal insulators, have excellent wear-resistant properties, and are stable in many corrosive environments. Fibers more than a centimeter in length are preferred for making thread, and these fibers are used for electrical insulation and other specialized uses. The large volume of short fibers produced is bound by inert media, such as portland cement (to produce roof and wall shingles) and vinyl plastics (to produce tiles).

Chrysotile asbestos apparently forms large deposits in only one way: by the near-surface hydration and alteration of the minerals in peridotite to form serpentine. The world's largest producer, Canada, obtains its supplies principally from a belt of serpentine running northeasterly across Quebec and extending into Vermont where there is a small-scale U.S. production. Although serpentines are not particularly uncommon rocks in mountain belts, asbestos resembles the scarce metals in its supplies, for only a tiny fraction of the serpentines have large, commercially exploitable deposits. The world's largest reserves of asbestos are in the hands of the three largest producing countries, Canada, the U.S.S.R., and the Republic of South Africa (Table 8-2).

Table 8-2   Producers of Asbestos, 1971*

Country	Production (metric tons)
CANADA	1.48×10^6
U.S.S.R.	1.15×10^6
REPUBLIC OF SOUTH AFRICA	0.32×10^6
PEOPLES REPUBLIC OF CHINA	0.16×10^6
ITALY	0.12×10^6
U.S.A.	0.12×10^6
WORLD TOTAL	3.58×10^6

* The largest producing countries own the largest reserves. Reliable reserve estimates have not been made public but reserves are believed to be large and adequate to meet rising demands for many generations.

(After U.S. Bureau of Mines.)

NONMETALLICS IN THE FUTURE

The future seems bright for most nonmetallic minerals. Reserves tend to be large and potential resources even larger. Some commodities such as building stone are so abundant that it is pointless to even attempt putting numbers on reserves. Viewing the resource abundance, some experts have suggested that nonmetallic minerals should be considered similar to abundant metals, that is, society should try to find ways of using nonmetallics to replace scarce metals and other commodities in short supply. Perhaps this will be possible, and if it is, the society that results will certainly be a very different one from today's. The nonmetallics seem to provide yet another example of abundant and underused resources awaiting innovative technological advances.

nine

water

. . . pure water is becoming a critical commodity whose abundance is about to set an upper limit of economic evolution in a few parts of the Nation and inevitably will do so rather widely within half a century or less. Prudence requires that the Nation learn to manage its water supplies boldly, imaginatively, and with utmost efficiency. Time in which to develop such competence is all too short. (A. M. Piper, U.S. Geological Survey, Water-Supply Paper #1797, 1965.)

Water is such a vital and essential resource that to attempt to assign it a cash value would be pointless; simply put, it is the most valuable of all our resources.

Although much water is locked in the minerals of the crust, it is the free water of the hydrosphere from which man must draw his resources. The total amount of water in the hydrosphere is estimated to be 1.36×10^9 cubic kilometers, or 1.36×10^{21} liters, but it is very unevenly distributed (Fig. 9-1), with 97.2 percent residing in the oceans and 2.15 percent of the remainder trapped in the polar icecaps and glaciers. These two largest water reservoirs are of principal use as transportation media, but both hold some potential for the future as local sources of drinking and irrigation water. The remaining 0.65 percent of the water in the hydrosphere is the fraction on which we now rely, and it would soon be consumed were it not for the well-known hydrologic cycle (Fig. 9-2). The solar-energy-driven cycle of evaporation plus transpiration followed by condensation, then precipitation, assures a continuous supply and makes water a renewable resource. Questions of water are therefore not only to do with abundance, but also with distribution and rates. Some areas are well supplied; others are water poor. More than any other factor, availability of water determines the ultimate population capacity of a geographic province.

	Location	Water volume (liters)	Percentage of total water
Surface water			
	Fresh-water lakes	125×10^{15}	.009
	Saline lakes and inland seas	104×10^{15}	.008
	Average in stream channels	1×10^{15}	.0001
Subsurface water			
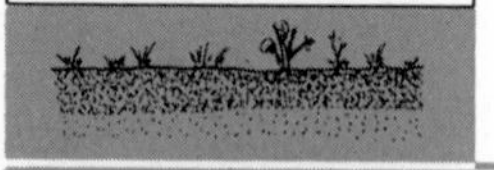	Vadose water (includes soil moisture)	67×10^{15}	.005
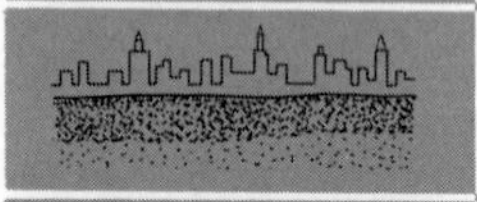	Ground water within depth of half a mile	$4,170 \times 10^{15}$	.31
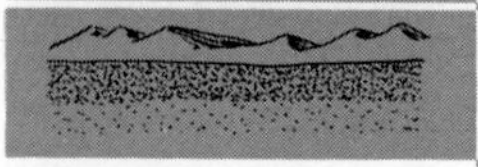	Ground water—deep lying	$4,170 \times 10^{15}$	.31
Other water locations			
	Icecaps and glaciers	$29,000 \times 10^{15}$	2.15
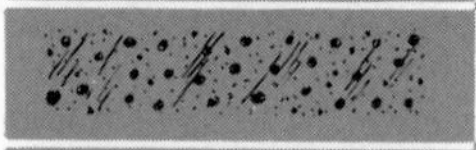	Atmosphere	13×10^{15}	.001
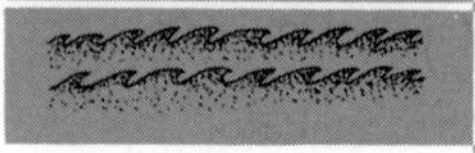	World ocean	$1,320,000 \times 10^{15}$	97.2

FIG. 9-1 Distribution of water in the hydrosphere. (After U.S. Geological Survey.)

DISTRIBUTION OF PRECIPITATED WATER

Precipitation around the world is distributed very unevenly. A belt of high rainfall straddles the equator and is flanked by two desert, or low-precipitation belts (Fig. 9-3). The country for which most information is available on the details of the hydrological cycle is the United States. Each year an average of 5.88×10^{15} liters of water falls as rain. The rainfall is very uneven, with precipitation varying by amounts of 20 times or more across large provinces (Fig. 9-3). The regional disparities that develop as a result of this are perhaps best demonstrated by the fact that the area east of the Mississippi receives 65 percent of the country's total rainfall (excluding Alaska and Hawaii), whereas the area west of the Mississippi receives only 35 percent.

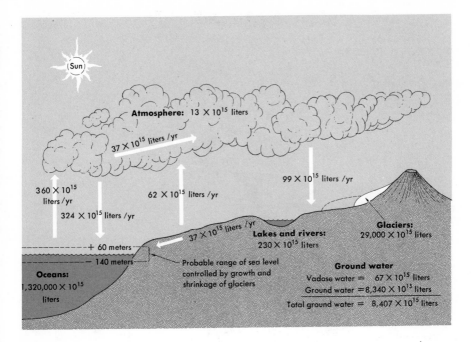

FIG. 9-2 The hydrosphere and the hydrologic cycle. (After A. L. Bloom, *The Surface of the Earth*, Prentice-Hall, 1969.)

In addition to geographic variations, there are marked temporal variations in precipitation. Seasonal variations are, of course, obvious to all, but there are also considerable long-term fluctuations in precipitation related to long-term weather cycles (Fig. 9-4). Although water is a renewable resource, the rate of renewal clearly is neither constant nor uniform across the country. Effective use of available water, therefore, requires both storage systems and efficient means of distributing the water to major consumption points. Both storage dams and distributions systems use mineral resources—for cements, pipes, pumps, and valves—and are another example of the influence of one resource we use on the production of others.

EVAPORATION AND TRANSPIRATION

Water will *evaporate* from any wetted surface. A significantly large fraction of the rainfall that falls on land is returned to the atmosphere in this fashion. In addition, water is assimilated by the root systems of growing plants and is later *transpired* from the leaf surfaces by a process essentially identical to evaporation. The two effects, evaporation and transpiration, cannot be individually discriminated for their effectiveness in returning rainfall to the atmosphere, but their sum contribution can be evaluated and is usually called the *evapotranspiration*

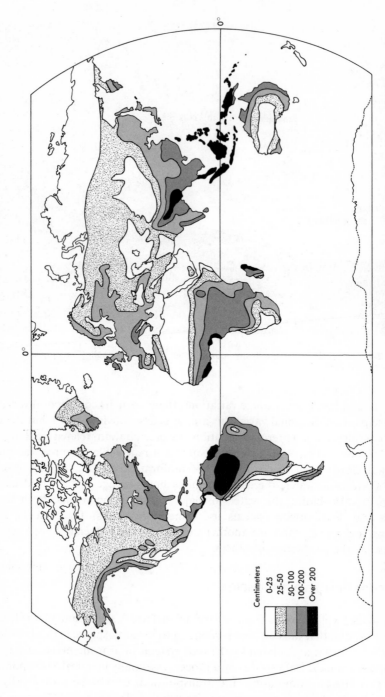

FIG. 9-3 Average precipitation around the world. Notice that the eastern half of North America enjoys a plentiful rainfall, while the western half is relatively dry.

Centimeters

0-25
25-50
50-100
100-200
Over 200

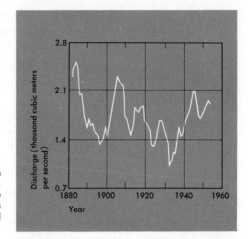

FIG. 9-4 Variations in the water discharge of the Mississippi River at Keokuk, Iowa, reflecting regular long-term variations in the rainfall and climate of the Mississippi River's drainage basin. (After U.S. Geological Survey.)

factor. The fraction of rain falling on the United States that is returned to the atmosphere by evapotranspiration, for example, is 70 percent; for the world as a whole, approximately 62 percent (see Fig. 9-2). In arid countries such as Australia the fraction is larger, and in less arid areas such as the United Kingdom it is lower. Water returned to the atmosphere by evapotranspiration is unavailable to man, except in the sense that useful plants may be grown in the place of useless ones. It cannot be trapped and redistributed for industrial or other purposes.

In regions of low rainfall, plant cover will develop to a point where all precipitation is used in evapotranspiration and none remains for stream flow. Seasonal rainfalls provide a qualifier for this statement because streams will flow even in the most arid areas during periods of maximum rainfall. In general, if the potential evapotranspiration—that which would result from the maximum plant cover a region could support under ideal circumstances—should exceed the precipitation, overland stream flow ceases. Conversely, if evapotranspiration is less than precipitation, runoff is generated.

The amount by which precipitation exceeds evapotranspiration is the perennial yield of stream flow water, and this is the usable fraction of rain and snowfall. Across the entire United States the water yield amounts to 30 percent of the total rainfall, or approximately 1.77×10^{15} liters per year. But when we consider the distribution of water deficiencies and surpluses we find that essentially the entire eastern half of the United States, together with a small region in the Pacific northwest, enjoys water surplus, while most of the country to west of the Mississippi is water-deficient and arid.

GROUND WATER

Although rainfall is the only water supply available on a long-term basis, there is a very important temporary reservoir available to man. Water in any form that occurs in the ground is commonly called *subsurface water*. It is one of the most important water resources, but unfortunately its rate of replenishment is often so slow that overly rapid withdrawal can cause serious local depletion. Indeed, in some areas the rate of renewal is so slow that subsurface water has to be considered a nonrenewable resource. Beneath the land areas of the world there is a zone where all rock pores and openings are saturated with water. This is *ground water*, and the upper surface of the saturated zone is the *water table*. The water table may lie at the surface, as in a lake or stream, or hundreds of meters below the surface; however, it is always present. Water below the water table is moving by seepage and slow flow to the sea (see Fig. 9-2); the ground water region can thus be considered analogous to a large but very slow river in which residence times for ground water to reach the sea may vary from hours to hundreds of thousands of years. The water table is not horizontal, but because of the varying resistance offered by rocks to the flow of ground water, it is irregular and tends to reflect the topography above (Fig. 9-5).

Above the water table is an unsaturated region of *vadose water*. Near the surface, where plant roots are abundant, is the region of *soil water*, in which water is moving neither up nor down, but adhering to the surface of mineral grains. Somewhat deeper, but still in the unsaturated zone, vadose water is slowly seeping down to join the body of ground water below. Vadose waters cannot be considered direct resources as can ground waters, but it is the vadose waters that serve to replenish—albeit slowly—any withdrawal from the ground water zone.

The amount of ground water is huge (Fig. 9-1), an estimated 3,000 times larger than the volume of water in all the rivers at any given time. The major problems in its exploitation are threefold. First, where rock porosity is very low and permeability poor, flow into wells is so slow that water cannot be removed at a worthwhile rate. Adequate recovery thus requires a suitable aquifer, or water carrier, as a supplier. Second, the rate of replenishment is slow because much of the supply—rainfall—runs off in rivers. It has been estimated, for example, that the total ground water resources in the United States to a depth of 750 meters would take an average of 150 years to be recharged if they were all removed, although some recharging areas would of course be slower than others. If

FIG. 9-5   Relation between subsurface waters, water table, and topography.

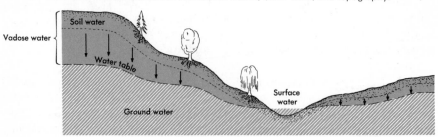

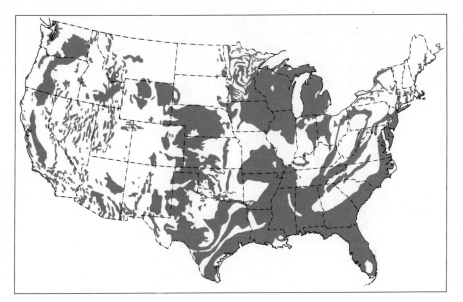

FIG. 9-6   Regions of the United States underlain by aquifers capable of yielding water containing 0.2 percent dissolved solids or less at rates of 190 liters per minute from individual wells. (After H. E. Thomas, 1955.)

water is pumped from the ground at a rate greater than that of replenishment, it is essentially being mined. This problem has now become acute in parts of the arid southwestern United States, where withdrawal at rates of up to a hundred times greater than those of replenishment has unfortunately been practiced in some of the richer ground water zones; replenishment, assuming cessation of all pumping, would take periods of up to 100 years or more. Finally, there is the problem of water quality. As ground water moves through the rocks, it dissolves the more soluble constituents. The problem varies with host rock, water depth, flow rates, and other factors, but in general a water with more than 0.05 percent (500 parts per million) of dissolved salts is unsuitable for human consumption, and one with more than 0.2 percent dissolved salts is unsuitable for almost all other uses; however, waters as saline as 1.0 percent can be used for some special purposes. The portion of the United States underlain by aquifers that will yield good-quality water into wells at flow rates of 190 liters per minute or more is extensive (Fig. 9-6), and wise development and utilization of these resources is now the prime concern of many able scientists and engineers.

RATES OF USE AND RESERVES

In 1965 a water specialist for the U.S. Department of the Interior cogently stated that ". . . all parts of the [United States] either have or will have water problems. The well-watered Eastern and Southern States are beginning to share a concern about water that has been felt in the arid West since its settlement.

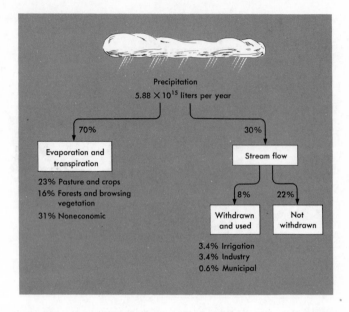

FIG. 9-7 Generalized distribution and use of the annual precipitation in the United States. (After A. Wolman, Water Resources, Pub. 1000-B, Committee on Natural Resources, National Academy of Sciences, National Research Council, Washington, D.C., 1962.)

As industrial development and urbanization expand in the East, it is becoming more apparent each year that lack of water may deter growth unless action is taken to assure a continued supply." Similar kinds of statements can be made for many parts of the world.

The current use of rainfall across the United States is summarized in Fig. 9-7. Approximately 8 percent of the rain falling on the United States, amounting to 0.42×10^{15} liters, is withdrawn for use in the categories shown. Of course, not all of the water withdrawn for use is consumed in a single use cycle. Water used to cool industrial machines, for example, can be returned to streams and reused. Irrigation and other agricultural water, however, is very largely consumed by evapotranspiration and hence is lost; it has been estimated that 25 percent of all water now used in the United States is so consumed and thus unavailable for further use.

Taking account of the fraction of water consumed, and projecting future population growth, Wolman has estimated that by 1980 the use of water must rise to 51 percent of the stream flow; by the year 2000, to 81 percent.

Steam flow varies with time, but man's water demands are constant. We are forced to build suitable damming, storage, and reticulation schemes so that excess flows of good seasons can be conserved to supplement diminished flows of poor seasons. In so doing, of course, large open water bodies such as reservoirs and lakes are created, and evaporation loss inevitably rises. Though experts differ, it has been optimistically stated that as much as 90 percent of the stream flow water may eventually be utilized. Even this optimistic estimate, which is made under the assumption that the necessary engineering and political problems could all be overcome, is dangerously close to Wolman's predicted withdrawal needs by the year 2000. The sense of urgency expressed by Dr. Piper is very real.

Rainfall, stream flow, and ground water are reserves. The oceans and the polar icecaps are potential resources, but both present use problems of vast magnitude. Sea water must be desalinated (meaning rid of its salts), and polar ice, though fresh water, has somehow to be brought to consumption areas. Many desalination plants working on evaporation principles are already in operation, but the cost of such a plant and the power to drive it make desalinated water expensive—about 20 cents per 1,000 liters, which is ten times more expensive than water from even the most expensive dam and reservoir system. Use of polar ice has not yet been tried. One estimate of economic feasibility showed that a tugboat could pull an iceberg 250 meters thick by 3,000 meters long by 3,000 meters wide from Antarctica to the arid west coast of Australia for about $1 million. If the iceberg could be encased in a plastic sheet to prevent evaporation and mixing with sea water, 790×10^9 liters of water could be delivered for a cost within a competitive range for economic use as agricultural water.

A problem faces use of water that does not afflict other resources—the problem of pollution. All of the projections and discussion made for water use are made under the assumption that adequate water purity is maintained. The seriousness of the problem of water pollution is self-evident. As the percentage of stream flow withdrawn for use rises, it becomes less and less feasible to use the same stream channels for effluent disposal, because a small quantity of polluted water can cause a whole stream to become polluted and hence unusable. The world's growing population is bringing with it a water problem fast enough, and to compound the problem by failure to control pollution, and hence to squander our most valuable natural resource, would be foolhardy in the extreme.

appendix

Table A-1 Units and Their Conversions

MULTIPLES AND SUBMULTIPLES

	Name	Common Prefixes
10^{12} = 1,000,000,000,000	trillion	
10^9 = 1,000,000,000	billion	giga
10^6 = 1,000,000	million	mega
10^3 = 1,000	thousand	kilo
10^2 = 100	hundred	hecto
10^1 = 10	ten	deka
10^{-1} = 0.1	tenth	deci
10^{-2} = 0.01	hundredth	centi
10^{-3} = 0.001	thousandth	milli
10^{-6} = 0.000001	millionth	micro

CONVERSIONS
BETWEEN COMMON UNITS OF MEASURE

Linear, Area, and Volume Measures

1 kilometer	= 0.6214 mile
1 meter	= 3.281 feet
1 centimeter	= 0.3937 inch
1 square kilometer	= 0.386 square mile
1 square meter	= 10.764 square feet

1 square centimeter = 0.155 square inch
1 cubic kilometer = 0.240 cubic mile
1 cubic meter   = 35.32 cubic feet
                   = 264.2 gallons (U.S.)
1 liter   = 0.264 gallon (U.S.)
1 barrel (oil)   = 42 gallons (U.S.)

Weight and Mass Measures

1 metric ton = 10^3 kilogram
1 long ton   = 2,240 pounds
1 short ton = 2,000 pounds
1 metric ton = 0.984 long ton
             = 1.102 short tons
1 kilogram   = 2.205 pounds

Energy and Power Measures

1 joule   = 0.239 calorie
1 calorie   = 3.9685×10^{-3} British thermal unit (Btu)
1 kilowatt hour = 10^3 watt hours
                = 3.6×10^6 joules
                = 3,413 Btu
1 watt   = 3.4129 Btu per hour
                = 1.341×10^{-3} horsepower
                = 1 joule per second
                = 14.34 calories per minute

Average Equivalents (as used in this book)

1 barrel oil weighs approximately 136.4 kilograms
1 barrel oil is equivalent to approximately 0.22 metric ton coal
1 barrel oil yields approximately 6.0×10^9 joules of energy
1 metric ton of coal yields approximately 27.2×10^9 joules of energy
1 barrel of cement weighs 170.5 kilograms

Table A-2   Principal Ore Minerals and Annual World Production (1973) of the Technologically Important Metals

I. THE GEOCHEMICALLY ABUNDANT METALS

Elements	World Production (metric tons)	Principal Ore Minerals
IRON	785,750,000 (iron ore; average about 60 percent Fe)	Magnetite, Fe_3O_4; hematite, Fe_2O_3; goethite, $HFeO_2$; siderite, $FeCO_3$
ALUMINUM	10,284,000 (metal)	Gibbsite, H_3AlO_3; diaspore, $HAlO_2$; boehmite, $HAlO_2$; kaolinite, $AL_2Si_2O_5(OH)_4$
TITANIUM	3,375,000 (ilmenite) 378,000 (rutile)	Ilmenite $FeTiO_3$; rutile, TiO_2
MANGANESE	20,672,000 (manganese ore; average about 45 percent Mn)	Pyrolusite, MnO_2; psilomelane, $BaMn_9O_{18} \cdot 2H_2O$; cryptomelane, KMn_8O_{16}; rhodocrosite, $MnCO_3$
MAGNESIUM	232,900 (metal) 11,730,000 (magnesite)	Magnesite, $MgCO_3$; dolomite, $CaMg(CO_3)_2$

II. THE GEOCHEMICALLY SCARCE METALS

A. Metals Commonly Forming Sulfide Minerals

Elements	World Production (metric tons)	Principal Ore Minerals
COPPER	6,044,000	Covellite, CuS; chalcocite. Cu_2S; digenite, Cu_9S_5; chalcopyrite, $CuFeS_2$; bornite, Cu_5FeS_4; tetrahedrite, $Cu_{12}Sb_4S_{13}$
ZINC	5,513,000	Sphalerite, ZnS
LEAD	3,403,000	Galena, PbS
NICKEL	640,400	Pentlandite $(Ni\ Fe)_9S_8$; garnierite, $H_4Ni_3Si_2O_9$
ANTIMONY	63,600	Stibnite, Sb_2S_3
MOLYBDENUM	78,470	Molybdenite, MoS_2
ARSENIC	greater than 40,000	Arsenopyrite, $FeAsS$; orpiment, As_2S_3; realgar, AsS
CADMIUM	15,490	Substitution for Zn in sphalerite
COBALT	22,730	Linnaeite, Co_3S_4; substitution for Fe in pyrite, FeS_2
MERCURY	10,540	Cinnabar, HgS; metacinnabar, HgS
SILVER	9,170	Acanthite, Ag_2S; substitution for Cu and Pb in their common ore minerals
BISMUTH	3,580	Bismuthinite, Bi_2S_3

B. Metals Commonly Found in the Native Form

Elements	World Production (metric tons)	Other Important Ore Minerals Besides the Native Elements
GOLD	1,446	Calaverite, $AuTe_2$; krennerite, (Au, Ag)Te_2; sylvanite, $AuAgTe_4$; petzite, $AuAg_3Te_2$
PLATINUM[b]	50.7	Sperrylite, $PtAs_2$; braggite, PtS_2; cooperite, PtS
PALLADIUM[b]	50.7	Arsenopalladinite, Pd_3As; michenerite, $PdBi_2$; froodite, $PdBi_2$
RHODIUM[b]	11.4	—
IRIDIUM[b]	7.6	—
RUTHENIUM[b]	5.1	Laurite, RuS_2
OSMIUM[b]	1.2	—

C. Metals Commonly Forming Oxygen-Containing Compounds

Elements	World Production (metric tons)	Principal Ore Minerals
CHROMIUM	6,291,000 (chromite)	Chromite, Fe_2CrO_4
TIN	233,200	Cassiterite, SnO_2
TUNGSTEN	36,600	Wolframite, $FeWO_4$; scheelite, $CaWO_4$
URANIUM[a]	20,100	Uraninite (pitchblende), UO_2 Carnotite, $K_2(UO_2)_2(VO_4)_2 \cdot 3H_2O$; substituting for Fe in magnetite, Fe_3O_4
VANADIUM	18,950	

Elements	World Production (metric tons)	Principal Ore Minerals
NIOBIUM AND TANTALUM[a]	10,550	Columbite, $FeNb_2O_6$; pyrochlore $NaCaNb_2O_6F$; tantalite, $FeTa_2O_6$
THORIUM	1,230	Monazite, a rare-earth phosphate containing thorium by atomic substitution
BERYLLIUM	250	Beryl, $Be_3Al_2(SiO_3)_6$

[a] Figures known for the Western-bloc nations only.
[b] Estimated by dividing the platinoid metal production of 126.8 metric tons in proportion to the relative metal abundances.

(After U.S. Bureau of Mines.)

suggestions for further reading

GENERAL

BROBST, D. A., AND PRATT, W. P., eds., *United States Mineral Resources*. U.S. Geological Survey Prof. Paper 820, 1973.

CLOUD, P. E., JR., *Resources and Man*. San Francisco: Freeman, 1969.

FLAWN, P. T., *Mineral Resources*, Chicago: Rand McNally, 1966.

LANDSBERG, H. H., *Natural Resources for U.S. Growth: A Look Ahead to the Year 2000*. Baltimore: Johns Hopkins, 1964.

LOVERING, T. S., *Minerals in World Affairs*. Englewood Cliffs, N.J.: Prentice-Hall, 1943.

McDIVITT, J. F., AND MANNERS, G., *Minerals and Men*. Baltimore: Johns Hopkins, 1974.

NATIONAL ACADEMY OF SCIENCES, *Mineral Resources and the Environment*, a report prepared by the Committee on Mineral Resources and the Environment, Feb. 1975.

SKINNER, B. J., AND TUREKIAN, K. K., *Man and the Ocean*. Englewood Cliffs, N.J.: Prentice-Hall, 1973.

THE STAFF, *Mineral Facts and Problems*. Bull. 630, U.S. Bureau of Mines, 1970.

THE STAFF, *Minerals Yearbook, Volumes I and II: Metals, Minerals and Fuels*. U.S. Bureau of Mines, published annually.

FUEL RESOURCES

AVERITT, P., *Coal Resources of the United States, Jan. 1, 1967*. U.S. Geological Survey Bull. 1275.

GRAY, T. J., AND GASHUS, O. K., *Tidal Power*. New York: Plenum Press, 1972.

LEVORSEN, A. I., *Geology of Petroleum*, 2nd ed. San Francisco: Freeman, 1967.

SCHURR, S. H., AND NETSCHERT, B. C., *Energy in the American Economy, 1850–1975. An Economic Study of its History and Prospects*. Baltimore: Johns Hopkins, 1960.

TIRATSOO, E. N., *Oil Fields of the World*. Beaconsfield, England: Scientific Press, 1973.

METAL DEPOSITS

BATEMAN, A. M., *The Formation of Mineral Deposits*. New York: Wiley, 1951.

DOUGLAS, R. J. W., ed., *Geology and Economic Minerals of Canada*. Economic Geology Report No. 1, Geological Survey of Canada, 1970.

PARK, C. F., JR., AND MACDIARMID, R. A., *Ore Deposits*, 2nd ed. San Francisco: Freeman, 1970.

ROUTHIER, P., *Les Gisements Métallifères, Tomes I et II*. Paris: Masson, 1963.

STANTON, R. L., *Ore Petrology*. New York: McGraw-Hill, 1972.

TITLEY, S. R., AND HICKS, C. L., *Geology of the Porphyry Copper Deposits of Southwestern North America*. Tucson: Univ. Arizona Press, 1966.

WARREN, K., *Mineral Resources*. New York, Wiley, 1973.

INDUSTRIAL MINERALS

BATES, R. L., *Geology of the Industrial Rocks and Minerals*. New York: Harper, 1960.

BORCHERT, H., AND MUIR, R. O., *Salt Deposits—The Origin, Metamorphism, and Deformation of Evaporites*. Princeton: Van Nostrand, 1964.

GILLSON, J. L., AND OTHERS, *Industrial Minerals and Rocks*. Seeley W. Mudd Series, 3rd ed. American Institute of Mining, Metallurgical and Petroleum Engineers, 1960.

WATER RESOURCES

DAVIS, S. N., AND DEWIEST, R. J. M., *Hydrogeology*. New York: Wiley, 1966.

DEWIEST, R. J. M., *Geohydrology*. New York: Wiley, 1965.

FOX, C. S., *The Geology of Water Supply*. London: London's Technical Press, 1949.

MCGUINNESS, C. L., *The Role of Ground Water in the National Water Situation*. U.S. Geological Survey, Water-Supply Paper No. 1800, 1963.

U.S. DEPARTMENT OF AGRICULTURE, *Water, The Yearbook of Agriculture*, 1955.

WALTON, W. C., *Groundwater Resource Evaluation*. New York: McGraw-Hill, 1970.

index

GEOLOGIC TIME SCALE
AND SOME IMPORTANT DATES IN THE FORMATION OF MINERAL RESOURCES

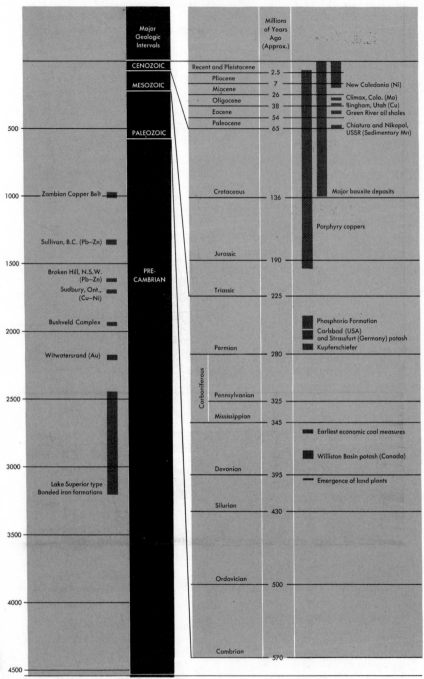

Major Geologic Intervals		Millions of Years Ago (Approx.)	
CENOZOIC	Recent and Pleistocene	2.5	
	Pliocene	7	New Caledonia (Ni)
MESOZOIC	Miocene	26	Climax, Colo. (Mo)
	Oligocene	38	Bingham, Utah (Cu)
	Eocene	54	Green River oil shales
	Paleocene	65	Chiatura and Nikopol, USSR (Sedimentary Mn)
PALEOZOIC			
	Cretaceous	136	Major bauxite deposits
PRE-CAMBRIAN			Porphyry coppers
	Jurassic	190	
	Triassic	225	
	Permian	280	Phosphoria Formation; Carlsbad (USA) and Strassfurt (Germany) potash; Kupferschiefer
	Pennsylvanian	325	
	Mississippian	345	
	Devonian	395	Earliest economic coal measures; Williston Basin potash (Canada); Emergence of land plants
	Silurian	430	
	Ordovician	500	
	Cambrian	570	

Zambian Copper Belt — 1000

Sullivan, B.C. (Pb–Zn) — 1500

Broken Hill, N.S.W. (Pb–Zn)
Sudbury, Ont., (Cu–Ni)

Bushveld Complex — 2000

Witwatersrand (Au)

2500

3000

Lake Superior type Banded iron formations

3500

4000

4500

Millions of Years

Formation of Earth's crust, about 4600 million years ago

Radiometric ages after Harland and others, 1964